Springer Theses

Recognizing Outstanding Ph.D. Research

Aims and Scope

The series "Springer Theses" brings together a selection of the very best Ph.D. theses from around the world and across the physical sciences. Nominated and endorsed by two recognized specialists, each published volume has been selected for its scientific excellence and the high impact of its contents for the pertinent field of research. For greater accessibility to non-specialists, the published versions include an extended introduction, as well as a foreword by the student's supervisor explaining the special relevance of the work for the field. As a whole, the series will provide a valuable resource both for newcomers to the research fields described, and for other scientists seeking detailed background information on special questions. Finally, it provides an accredited documentation of the valuable contributions made by today's younger generation of scientists.

Theses may be nominated for publication in this series by heads of department at internationally leading universities or institutes and should fulfill all of the following criteria

- They must be written in good English.
- The topic should fall within the confines of Chemistry, Physics, Earth Sciences, Engineering and related interdisciplinary fields such as Materials, Nanoscience, Chemical Engineering, Complex Systems and Biophysics.
- The work reported in the thesis must represent a significant scientific advance.
- If the thesis includes previously published material, permission to reproduce this must be gained from the respective copyright holder (a maximum 30% of the thesis should be a verbatim reproduction from the author's previous publications).
- They must have been examined and passed during the 12 months prior to nomination.
- Each thesis should include a foreword by the supervisor outlining the significance of its content.
- The theses should have a clearly defined structure including an introduction accessible to new PhD students and scientists not expert in the relevant field.

Indexed by zbMATH.

More information about this series at https://link.springer.com/bookseries/8790

Seiji Fujimoto

Demographics of the Cold Universe with ALMA

From Interstellar and Circumgalactic Media to Cosmic Structures

Doctoral Thesis accepted by University of Tokyo, Tokyo, Japan

Springer

Author
Seiji Fujimoto
Cosmic DAWN Center
Copenhagen, Denmark

Supervisor
Prof. Masami Ouchi
Department of Astronomy
University of Tokyo
Tokyo, Japan

ISSN 2190-5053 ISSN 2190-5061 (electronic)
Springer Theses
ISBN 978-981-16-4978-3 ISBN 978-981-16-4979-0 (eBook)
https://doi.org/10.1007/978-981-16-4979-0

This Springer imprint is published by the registered company Springer Nature Singapore Pte Ltd.
The registered company address is: 152 Beach Road, #21-01/04 Gateway East, Singapore 189721, Singapore

Supervisor's Foreword

In the sixteenth century, Thomas Digges recognized the strange fact that the night sky was dark. Now it is known as the Olbers' paradox, an argument that the darkness of the night sky is contradictory to the infinite and eternal static universe. After recognitions of the importance of the night sky brightness, referred to as the background light, in the cosmological context, the background light was detected in γ-ray to radio wavelengths in the second half of the twentieth century. Among the background light, the Cosmic Infrared Background (CIB) was accurately measured by Cosmic Background Explorer (COBE), and revealed that the total energy of CIB emission was as high as one of the cosmic optical (stellar) background lights.

The major origin of CIB was unclear. In the 1980s, the Infrared Astronomy Satellite (IRAS) has identified a large population of local dusty starbursts whose interstellar cold dust, heated by massive stars, emits infrared-continuum radiation that contributes to the CIB. Subsequently, submillimeter observations have detected high-redshift dusty starbursts of high star-formation rates (SFRs : 1000 M_e yr^{-1}), a.k.a. submillimeter galaxies, with very bright ($10^{12-14} L_e$) infrared emission that is redshifted to the submillimeter band in the observed frame. While the sources of infrared emission were found in the local and high-redshift universe, only a fraction of the CIB were resolved, due to the limited sensitivities and statistics.

Such observations also revealed rest-frame infrared continua and emission lines in starburst galaxies, especially [C II] 158μm, radiated from cold gas of the interstellar medium (ISM) in the local and high-redshift universe. The [C II] emission is one of the strongest emission lines of galaxies in the rest-frame infrared band, where the [C II] luminosity is as bright as 1% of the total infrared luminosity in galaxies. Although the bright [C II] line and the continuum are good probes for spatial distributions of the cold ISM, the spatial distributions of [CII] and continuum were known only in luminous infrared galaxies in the local universe until recently.

The following Ph.D. thesis written by Seiji Fujimoto addresses these important issues, exploiting observational data of the state-of-the-art radio interferometer, the Atacama Large Millimeter/submillimeter Array (ALMA). Although the sensitivity of ALMA is extremely higher than those of the other existing radio facilities, the small area and frequency coverages of ALMA do not provide a statistical data set even in large observational programs approved by the early ALMA time allocation committees. Fujimoto's thesis uses ALMA archival data, in addition to his original data, to make the largest ALMA samples ever obtained, pioneering statistical studies of cold dust and gas in high-redshift star-forming galaxies (their SFRs down to a few M_e yr^{-1}) with the ALMA archive. On the basis of comprehensive analyses of the ALMA data, Fujimoto presents three major conclusions on the scales over the galaxy ISM to the cosmic structures that are summarized below in an order of scientific relevance.

First, dust-continuum radiation is confined in a compact region of star-forming galaxies with a disk-like morphology of Sérsic index $n \sim 1$. Although the dusty star-formation region is compact, the spatial distributions of star formation traced by ultraviolet radiation of massive stars and infrared radiation of heated dust suggest that a stellar profile of elliptical galaxy with $n : 4$ cannot be produced by the star-forming activities, but dynamical processes taken place after the star formation.

Second, the sky surface density of dust-continuum sources of star-forming galaxies resolved by this thesis is high enough to explain the CIB. In other words, the major origin of the CIB is not dust in the inter-galactic space, but in the compact regions of star-forming galaxies.

Third, the last and the most surprising result of this thesis is the first identification of a 10-kpc-sized [C II] emitting halo around high-redshift star-forming galaxies. Because carbon is not produced by the Big Bang nucleosynthesis, carbon ions emitting the [C II] line are probably remnants of early gas outflow from the high-redshift galaxies. No galaxy-formation models have so far predicted such a large halo of cold gas emitting [C II] in the circumgalactic medium. Fujimoto claims that the galaxy-formation models, to date, miss important processes such as strong feedback making a significant cold gas outflow.

The recognition of this thesis work led to Fujimoto taking a prestigious research position, DAWN Fellowship, in Copenhagen, Denmark, that allows him to conduct independent researches for innovative results. He is now extending his ALMA studies toward two directions as his new challenges, high-redshift galaxy halos and evolution of supermassive black holes, exploiting his experiences in the ALMA archival statistical studies presented in this thesis. This thesis is the first set of his innovative studies, and worth reading for those who aim to investigate the cold side of the universe and to carry out statistical work with big data in radio wavelengths.

Tokyo, Japan
February 2020

Masami Ouchi

List of Publication

Parts of this thesis have been published in the following journal articles:

- S. Fujimoto, M. Ouchi, A. Ferarra, A. Pallottini, R. J. Ivison, C. Behrens, S. Gallerani, S. Arata, H. Yajima, and K. Nagamine "First Identification of 10-kpc Scale [C II] Halo around Star-Forming Galaxies at $z = 5 - 7$", 2019, ApJ, 887, 107.
- S. Fujimoto, M. Ouchi, K. Kohno, Y. Yamaguchi, B. Hatsukade, Y. Ueda, T. Shibuya, S. Inoue, T. Oogi, S. Toft, C. Gomez-Guijarro, T. Wang, T. Nagao, I. Tanaka, Y. Ao, H. Umehata, Y. Taniguchi, K. Nakanishi, W. Rujopakarn, R. J. Ivison, W. Wang, M. Lee, K. Tadaki, Y. Tamura, and J. S. Dunlop "ALMA 26 arcmin^2 Survey of GOODS-S at One-millimeter (ASAGAO): Average Morphology of High-z Dusty Star-Forming Galaxies is an Exponential-Disk (n; 1)", 2018, ApJ, 861, 7.
- S. Fujimoto, M. Ouchi, T. Shibuya, H. Nagai, "Demonstrating a New Census of Infrared Galaxies with ALMA (DANCING—ALMA) I. FIR Size and Luminosity Relation at $z = 0 - 6$ Revealed with 1034 ALMA Sources", 2017, ApJ, 850, 83.
- S. Fujimoto, M. Ouchi, Y. Ono, T Shibuya, M. Ishigaki, H. Nagai, R. Momose "ALMA Census of Faint 1.2 mm Sources down to :0.02 mJy: Extragalactic Background Light and Dust-Poor High-z Galaxies", 2016, ApJS, 222, 1.

Acknowledgements

Special thanks to my thesis supervisor Masami Ouchi for his invaluable advice, encouragement, and support during the period of my Ph.D. I also deeply thank Ph.D thesis committee members: Satoru Iguchi, Ryohei Kawabe (Chair), Kentaro Motohara, Toshikazu Shigeyama, and Hidenobu Yajima for their constructive comments and suggestion. I am grateful to Kotaro Kohno and Yoichi Tamura for leading me to an exciting field of radio astronomy. I appreciate the group members including the previous members: Yoshiaki Ono, Takatoshi Shibuya, Jun Toshikawa, Ken Mawatari, Rieko Momose, Mariko Kubo, Akira Konno, Yoshiaki Naito, Masafumi Ishigaki, Hiroko Tamazawa, Yuichi Harikane, Takashi Kojima, Shiro Mukae, Yuma Sugahara, Ryo Higuchi, Haibin Zhang, Miftaful Hilmi, Shotaro Kikuchihara, Ryohei Itoh, Ryota Kakuma, Karin Shimodate, Nao Sakai, Yuki Isobe, and Takako Idomura for daily discussions, chats, taking lunch/dinner, playing tennis, and going to the hot spring that motivate me to enjoy the daily research life in Kashiwa. I thank Yuki Yamaguchi, Haruka Kusakabe, and Taichi Uyama for studying and encouraging together from undergraduate in the Department of Astronomy.

I would like to express my gratitude to Hiroshi Nagai, Bunyo Hatsukade, Yuichi Matsuda, Tadayuki Kodama, Soh Ikarashi, Hideki Umehata, Izumi Takuma, Koichiro Nakanishi, Daniel Espada, Kazuya Saigo, Daisuke Iono, Fumi Egusa, Junko Ueda, Toshiki Saito, Akifumi Seko, Akio Taniguchi, and Minju Lee for giving me helpful supports, advice, and fruitful discussions as the experts of the extragalactic studies with the submillimeter/millimeter facilities including ALMA. I would also appreciate the helpful feedbacks and advice offered by Tohru Nagao, Akio Inoue, Koji Ohta, Yoshihiro Ueda, Hidenobu Yajima, Masayuki Umemura, Yoshiaki Taniguchi, Ken-ichi Tadaki, Takuya Hashimoto, Ryota Kawamata, Masafumi Onoue, Kyoko Onishi, and Ryosuke Goto.

I am indebted to the Italian colleagues of Andrea Ferrara, Andrea Pallottini, Christoph Behrens, Simona Gallerani, and Graziano Ucci for supporting my research stay in Pisa, as well as other foreign colleagues of Sune Toft, Olivier Le Fèvre, David Elbaz, Raffaella Schneider, J. S. Dunlop, R. J. Ivison, Nick Scoville, Ian Smail, Dan Coe, and Fabian Walter for hosting me in the seminars. I would also thank all the collaborators all over the world including Tao Wang, Ichi Tanaka, Yiping Ao, Wiphu

Rujopakarn, Wei-hao Wang, Walter Gear, Carlos Góme-Guijarro, Darko Donevski, Scott Chapman, James Simpson, Clements Dave, Marcin Sawicki, Chien-Hsiu Lee, Roland Bacon, Jeremy Blaizot, and Floriane Leclercq.

I appreciate the support of the staff at the ALMA Regional Center. This thesis makes use of the following ALMA data: ADS/JAO.ALMA #2011.0.00115.S, #2011.0.00232.S, #2011.0.00319.S, #2011.0.00648.S, #2011.0.00767.S, #2012.1.00076.S, #2012.1.00602.S, #2012.1.00719.S, #2012.1.00323.S, #2012.1.00536.S, #2012.1.00610.S, #2012.1.00934.S, #2012.1.00953.S, #2011.0.00097.S, #2011.0.00294.S, #2012.1.00076.S, #2012.1.00090.S, #2012.1.00245.S, #2012.1.00307.S, #2012.1.00326.S, #2012.1.00523.S, #2012.1.00756.S, #2012.1.00775.S, #2012.1.00869.S, #2012.1.00979.S, #2012.1.00983.S, #2013.1.00034.S, #2013.1.00118.S, #2013.1.00151.S, #2013.1.00173.S, #2013.1.00718.S, #2013.1.00884.S, #2013.1.00205.S, #2013.1.00208.S, #2013.1.00566.S, #2013.1.00668.S, #2013.1.00781.S, #2013.1.00999.S, #2013.1.01292.S, #2015.1.00137.S, #2015.1.00540.S, #2015.1.00664.S, #2015.1.01495.S, #2015.1.00098.S #2013.1.00815.S, #2015.1.00834.S, #2015.1.01111.S, #2015.1.01105.S, #2016.1.01240.S, #2012.1.00523.S, and #2012.1.00602.S. ALMA is a partnership of ESO (representing its member states), NSF (USA) and NINS (Japan), together with NRC (Canada) and NSC and ASIAA (Taiwan), in cooperation with the Republic of Chile. The Joint ALMA Observatory is operated by ESO, AUI/NRAO, and NAOJ. This work was supported by the World Premier International Research Center Initiative, MEXT, Japan, and KAKENHI 16J02344, (A) 23244025, (A) 15H02064, (A) 17H01110, (A) 17H01111, (A) 17H01114, and (S) 17H06130 Grant-in-Aid for Scientific Research through the Japan Society for the Promotion of Science, Graduate Research Abroad in Science Program 2017, Hayakawa Satio Fund awarded by the Astronomical Society of Japan (cycle 89, 95, 106), and ALMA Japan research Grant of NAOJ Chile Observatory (ID145, 164, 179, 197).

Contents

Chapter 1
Introduction

1.1 Hidden, Cold Side of the Universe

There are three major energies coming from the universe as the cosmic background light (Fig. 1.1): cosmic optical (COB), infrared (CIB), and microwave background light (CMB). The COB and CMB are explained by the stellar continuum of galaxies and the leftover radiation from the Big Bang, respectively. On the other hand, the origin of the CIB has not been fully understood since the CIB was first identified by the *Cosmic Background Explorer* (*COBE*) satellite [1–5].

The universe reserves a large amount of the cold gas that forms stars. These stars produce the dust when they die via the supernova explosions, and the dust soon becomes cold through widely distributed in the cold gas. The ultra-violet (UV) emission originated from the stars is absorbed by the cold dust and re-emitted as the thermal infrared emission. Because the CIB is thought to be contributed by the cold dust emission, the CIB shows another side of the universe obscured by the cold dust (i.e., Cold Universe). It is thus essential to unveil the Cold Universe: where and how much amount of the cold dust emission is originated in the universe. Furthermore, it is also required to reveal the physical properties of the individual cold dust emission to understand what kind of roles the dusty obscured star-forming activity plays in the galaxy formation and evolution.

1.2 Hierarchical Structure of the Universe

Where does the cold dust emission come from ? Galaxies are thought to be formed in their dark matter halos in the frameworks of Λ cold dark matter (ΛCDM) structure formation models. These galaxies are evolved in the interplay with the surrounding medium via gas inflow [e.g., 6] and outflow [e.g., 7] that regulate star-formation by gas-supply and feedback processes [8]. Figure 1.2 presents the numerical simulation results for the gas density in a $100 \times 100 \times 20$ comoving Mpc box. In Fig. 1.2, while the

S. Fujimoto, *Demographics of the Cold Universe with ALMA*, Springer Theses,
https://doi.org/10.1007/978-981-16-4979-0_1

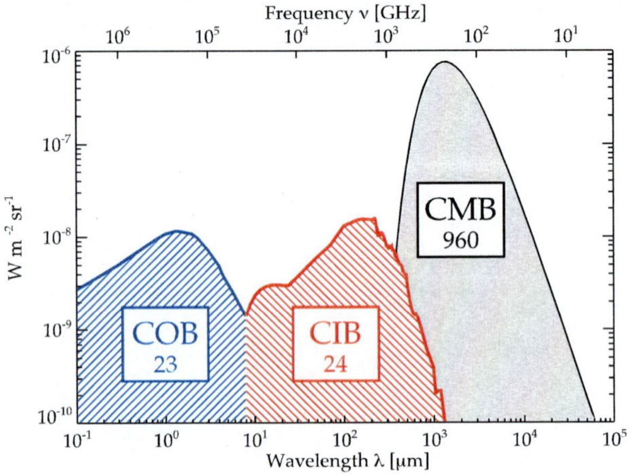

Fig. 1.1 Schematic Spectral Energy Distributions of the most important (by intensity) backgrounds in the universe, and their approximate brightness in nW m^{-2} sr^{-1} written in the boxes (Credit: Dole et al., A&A, 451, 417, 2006, reproduced with permission ©ESO.)

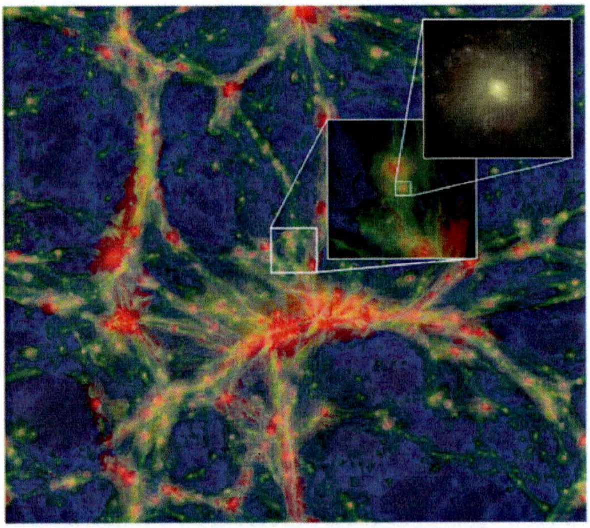

Fig. 1.2 A 100×100×20 comoving Mpc (cMpc) slice through the EAGLE simulation, illustrating the dynamic range attainable with state-of-the-art SPH simulations (Fig. 1 of [9], MNRAS, 446, 521, reproduced by permission of Oxford University Press). The intensity represents the gas density while the color indicates the gas temperatures (blue=green-red from cooler to hotter). The inset shows a region 10 cMpc and 60 ckpc on a side. The zoom in to an individual galaxy with stellar mass $3\times10^{10} M_\odot$ shows the optical band stellar light

galaxy is formed in the local gas density peak with the interstellar medium (ISM), the large amount of the gas is still widely distributed in the scales of the circumgalactic medium (CGM) and the cosmic structure. Since the cold gas and dust generally coexist through the interplay, the cold dust emission is expected to be originated in the scales of the CGM and the cosmic structure as well as the ISM.

To comprehensively understand the obscured side of the galaxy formation and evolution in this hierarchical structure of the Cold Universe, it is thus important to fully resolve the cold dust emission in the ISM, the CGM, and the cosmic structure instead of focusing on a part of them.

1.2.1 Inter-Stellar Medium Scale

Size and morphology of high-redshift galaxies are the key to understand the galaxy formation and evolution. In particular, the effective radius R_e and the Sérsic index n [10, 11] are key quantities to evaluate the size and morphological properties.

In the rest-frame ultra-violet (UV) and optical wavelength regimes, the *Hubble Space Telescope* (HST) has revealed the R_e and n properties for the high-redshift galaxies up to $z \sim 6 - 10$ [e.g., 12–21]. These HST studies show that quiescent and star-forming galaxies are represented by a spheroidal morphology with a Sérsic index in the rest-frame optical wavelength $n_{opt} \sim 4$ and an exponential-disk morphology with $n_{opt} \sim 1$, respectively. Moreover, the quiescent galaxies are more compact than the star-forming galaxies at a given stellar mass. These HST results indicate that the transformation in both size and morphology might be attributed to the evolution from star-forming to quiescent galaxy (Fig. 1.3). However, the actively star-forming regions obscured by dust cannot be unveiled with the rest-frame UV and optical studies. The stars formed in the dusty starburst may eventually dominate over the stars already present in the host galaxies due to an extreme star formation rates in these dusty star-forming galaxies. Studies of the size and morphology in the rest-frame far-infrared (FIR) wavelength are thus essential to comprehensively understand the evolutionary connections from the high-z star-forming to compact quiescent galaxies.

Recent ALMA observations enable us to measure n and R_e in the rest-frame FIR wavelength, n_{FIR} and $R_{e,FIR}$, due to its high sensitivity and angular resolution. For bright submillimeter galaxies (SMGs; $S_{1mm} \gtrsim 1$ mJy), recent ALMA studies report that SMGs have the exponential-disk morphology with $n_{FIR} = 0.9 \pm 0.2$ [22], and $R_{e,FIR}$ of $\sim 1 - 2$ kpc [e.g., 22–25] smaller than R_e in the rest-frame optical wavelength $R_{e,opt}$ of $\sim 3 - 4$ kpc [e.g., 25–28]. For faint SMGs ($S_{1mm} < 1$ mJy), although there are several attempts to estimate $R_{e,FIR}$ by using deep ALMA imaging [25, 29, 30], large uncertainties still remain due to the small number statistics and observational challenges.

There are two major observational challenges to measure n_{FIR} and $R_{e,FIR}$ for faint SMGs. First is the sensitivity. To perform a secure profile fitting, high signal-to-noise (S/N) level is required. Extremely deep observations are thus needed. Second is the angular resolution. The rest-frame UV and optical studies report a positive correlation

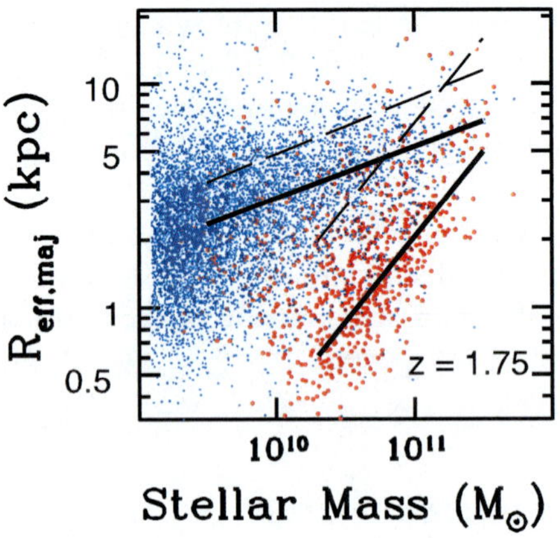

Fig. 1.3 Size-stellar mass distribution of late- (blue) and early-type (red) galaxies at $1.5 < z < 2$ (Fig. 5 of [17], reproduced by permission of the AAS). The lines indicate model fits to the early- and late-type galaxies. The dashed lines represent the model fits to the galaxies at redshifts $0 < z < 0.5$. The solid lines represent fits to the higher-redshift samples. The mass ranges used in the fits are indicated by the extent of the lines in the horizontal direction

between the galaxy size and luminosity [e.g., 17, 18], which suggests that faint SMGs are more compact than bright SMGs. To resolve the compact objects, we need high angular resolution.

1.2.2 Circum-galactic Medium Scale

In the early universe, star-forming galaxies are known to be surrounded by the CGM-scale neutral hydrogen, which emits in the Lyα line, giving rise to Lyα haloes [Fig. 1.4; e.g., 31, 32]. After they form, from the primordial (H+He) gas, the first stellar populations in the earliest galaxies produce heavier elements. Theoretical models predict these are expelled into the CGM via powerful outflows driven by active galactic nuclei and/or supernova explosions [33, 34]. The metal-enrichment in the primordial CGM is thus a powerful probe for the outflow activities in the early galaxies.

It is known that rare, massive galaxies—formed via mergers [35] and possibly cold-stream accretion [36], with an abundance of massive stars [37]—drive outflows throughout cosmic time, forming the metal-enriched CGM [38–41]. However, there has hitherto been no evidence that gas is expelled from primaeval galaxies with more modest star-formation rates (SFRs), which dominate the earliest period of cosmic star-formation history. Any expelled gas would be much more diffuse than the interstellar medium and there is thought to be insufficient energy available to drive outflows at low SFRs [42, 43].

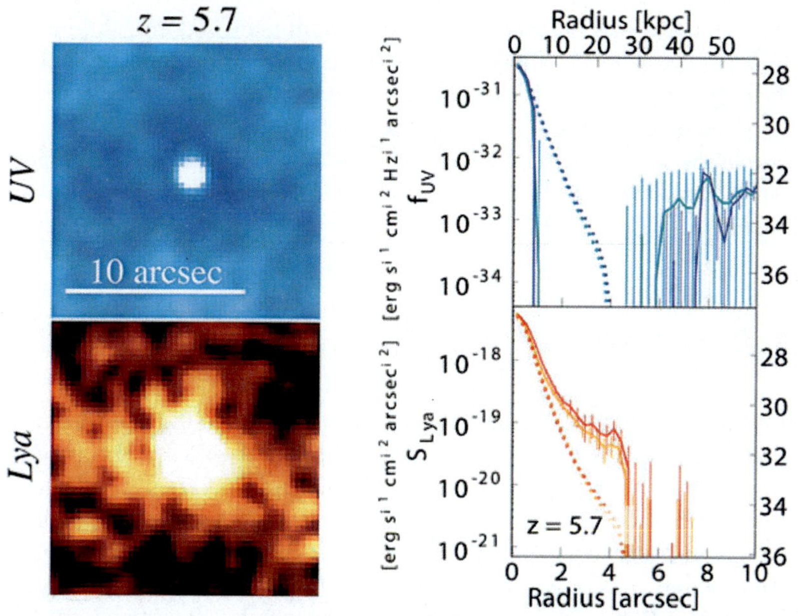

Fig. 1.4 **(Left)** Composite rest-frame continuum (top) and Lyα images of the $z = 5.7$ Lyman-alpha emitters (LAEs) produced by the mean-combined method. **(Right)** Radial surface brightness profiles of composite images of rest-frame UV continuum (top) and Lyα (bottom) from the $z = 5.7$ LAEs (solid lines) and PSFs (dotted lines). These figures are reproduced from Fig. 2 and 3 of [31], MNRAS, 442, 110, by permission of Oxford University Press on behalf of the sponsoring Monthly Notices of the Royal Astronomical Society

ALMA has opened our views to the metal-enriched medium in the early universe, providing the secure detections such as the dust continuum and the [CII]-158 μm line emission even up to $z \sim 7$ [e.g., 44–52]. There have also been several attempts to measure the size and morphology in the rest-frame FIR continuum and the [CII] line for such high redshift galaxies at $z \sim 5$–7, where [53] report that the effective radius of the [CII] line-emitting region is larger than that of the rest-frame UV region. However, large uncertainties still remain due to the small number statistics and observational challenges.

One critical challenge is the flux recovery for the extended emission with ALMA. In the interferometric observations, it is required to sample enough coverage of the uv-visibility, especially in the short baselines, to recover the flux of the extended emission, while the general ALMA [C II] observations for the high-z star-forming galaxies are carried out under the assumption that the [C II]-emitting regions are distributed in the ISM scale. If the [C II] line emission extends over the ISM scale, such extended [C II] emission might be missed in the previous ALMA observations. Another critical challenge is sensitivity. The recent ALMA studies show that signal-to-noise ratio (S/N) > 10 is needed to obtain reliable size measurement results both on the image-based and visibility-based analyses [e.g., 23, 24], while the majority of the previous ALMA detections of the dust continuum and the [CII] line from $z \sim$ 5–7 galaxies show the S/N less than 10. If the S/N level is poor, noise fluctuations significantly affect the profile fitting results. Moreover, [22] show that the combination of the original smoothed galaxy profile and the noise fluctuations can make the morphology more clumpy. To obtain the reliable size and morphological results without the missing flux, extensively deep and good uv-sampling observations are thus required.

1.2.3 Cosmic Structure Scale

In the cosmic structure scale, the origin of the CIB is one of the biggest questions. To reveal the origin of the CIB, first we need to detect individual galaxies and evaluate their contribution to the CIB. The most advantageous regimes are the submm/mm wavelengths due to the negative k-correction. In previous blank field observations with single-dish telescopes, it has revealed that the bright dusty high-z objects of submm galaxies [SMGs, $S_{1mm} \gtrsim 1$ mJy; 54] are not major contributors of the CIB. SCUBA and LABOCA observations have shown that the 20–40% of CIB is resolved at submm wavelength [e.g., 55–59]. Similar results have been obtained by AzTEC observations that they have resolved 10–20% of the CIB at mm wavelength [e.g., 60–62]. These observations suggest that there should exist high-z populations different from SMGs, and that such populations are major CIB contributors.

The contribution from the faint submm/mm sources can be now addressed by ALMA, owing to its capabilities of high angular resolution and sensitivity. In other words, recent ALMA observations allow us to study the faint submm/mm sources with the negligibly small uncertainties of the source confusion and blending which The first ALMA mm search for faint sources below $S_{1mm} \sim 1.0$ mJy with no lensing effect has been conducted by [63]. The authors resolve ~80% of the CIB down to $S_{1mm} = 0.1$ mJy in 20 targets residing in one blank field of the Subaru/XMM-Newton Deep Survey [SXDS; 64]. However, this result would include unknown effects of cosmic variance, because of the single field observations. Subsequently, [65, 66] have claimed the CIB resolved a fraction of ~60% that is smaller than the one of [63], utilizing multi-field deep ALMA 1mm maps of Bands 6 and 7 down to the flux limit of $S_{1mm} = 0.1$ mJy, where the uncertainties from cosmic variance are probably reduced with the multi-field data. As a brief summary of these ALMA studies, only about a half of the CIB has been resolved with the present ALMA flux limit of $S_{1mm} = 0.1$ mJy (Fig. 1.5), including the effects of the cosmic variance. No one still knows whether the remaining half of the CIB is resolved by the very faint ($S_{1mm} < 0.1$ mJy), individual submm/mm sources, or by the faint diffuse emission, extended over the CGM and cosmic structure scales, that has been missed in the previous studies.

There is another remaining issue about the origin of the CIB. Although these studies newly identify faint ALMA sources ($S_{1mm} < 1.0$ mJy) and their contribution to the CIB nearly the half, these studies do not clearly answer to the question about the

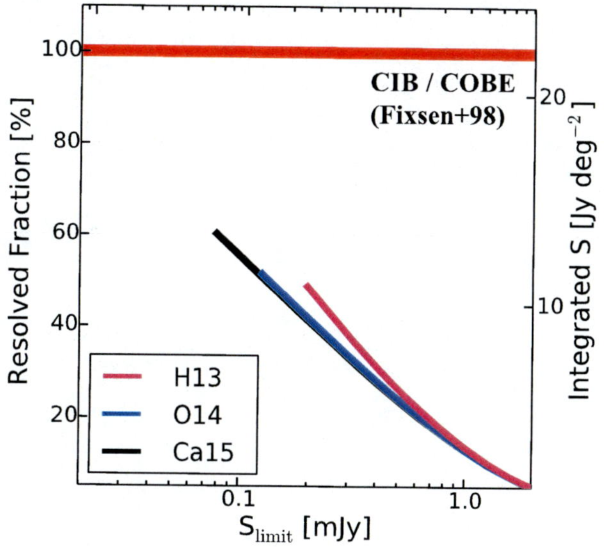

Fig. 1.5 Resolved fraction of the CIB with previous ALMA observations against the flux limit ($S_{\rm limit}$) at 1.2 mm. The magenta, blue, and black lines present the results of H13 [63], O14 [65], and Ca15 [66], respectively. These lines are obtained by re-fitting the Schechter functions to the data points of each literature and [62] for the bright end. The red line shows the CIB value estimated from the *COBE* observations [2]

connection between the faint ALMA sources and optically selected high-z galaxies. Recent observations with *Herschel* have revealed that typical UV-selected galaxies such as Lyman-break galaxies (LBGs) have a median total (8–1000 μm) luminosity of $L_{\rm IR} \simeq 2.2 \times 10^{11} L_\odot$ [67–69]. The stacking analysis of *Herschel* and ALMA data has also shown that K-selected galaxies, including star-forming BzK galaxies (sBzKs), have total IR luminosities of $L_{\rm IR} = (5 - 11) \times 10^{11} L_\odot$ [70]. These ranges of the IR luminosities of LBGs and sBzKs correspond to the mm flux of $S_{\rm 1mm} \sim 0.1 - 1$ mJy if we assume a modified black body with $\beta_{\rm d} = 1.8$, dust temperature $T_{\rm d} = 35$K, and a source redshift $z = 2.5$. These sources could be mm counterparts of the optically selected galaxies, because the mm flux range shown in the stacking analyses is similar to that of the faint ALMA sources.

There is another approach, focusing on the dark matter halo property, to characterize the faint ALMA sources. The clustering analysis is a powerful tool for understanding the connection between various galaxy populations, since the spatial distribution of galaxies is related to the underlying distribution of dark haloes in the standard scenario of structure formation in ΛCDM universe. Clustering analyses have been carried out over the past decade for bright SMGs observed with single-dish telescopes, and concluded that the clustering amplitudes, or galaxy biases, are large, and that the hosting dark halo masses of the bright SMGs are estimated to be $\sim 10^{13} M_\odot$ ([59, 71–74], cf. [75]). On the other hand, however, the dark halo properties of the faint ALMA sources are poorly known, except for the result of [65] who have obtained, for the first time, a meaningful constraint on a galaxy bias, $b_{\rm g} \leq 4$, for faint ALMA sources. In this way, the faint ALMA sources are not well studied. To understand the physical origin of faint ALMA sources, one should study faint ALMA sources with a more complete data set on the basis of individual sources as well as statistics.

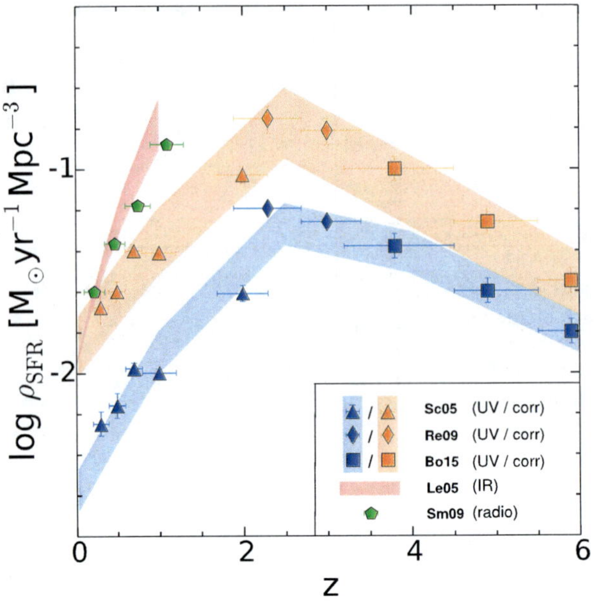

Fig. 1.6 Cosmic SFR density (ρ_{SFR}) measurements in previous optical-radio observations. The blue symbols are the SFR density measurements from the optical observations ($SFRD_{UV}$) in [76–78], referred to as "Sc05", "Re09", and "Bo15", respectively. The orange symbols are the SFR density values after dust correction ($SFRD_{corr}$) from the effects of dust extinction using the observed UV slopes β [79] and the IRX–β relationship [80]. The blue and orange regions are the uncertainties of the $SFRD_{UV}$ and $SFRD_{corr}$ values that are presented in Fig. 18 of [78]. The light-red region and green pentagon show the SFR density measurements from the direct IR observations (SFR_{IR}) in [81, 82], referred to as "Le05" and "Sm09", respectively. The SFR_{IR} measurements have been limited at $z \lesssim 1$

1.3 Cosmic Star-Formation Rate Density

As summarized in Sect. 1.2.1–1.2.3, no one still knows where and how much amount of the star-forming activity is obscured by dust in the universe. This indicates that the current understanding of the cosmic SFR density is incomplete. Figure 1.6 shows the current results of the cosmic SFR density measurements. The light red and blue shades are estimated from the rest-frame UV galaxies identified in the optical-NIR surveys with and without the dust correction, respectively, and the light red shade measurement is usually used to discuss the total (= dust obscured and un-obscured) cosmic SFR density. However, if the previous optical-near infrared (NIR) surveys have missed a number of the dusty obscured star-formation activities in the scales of the ISM, the CGM, and the cosmic structure, the SFR density in the previous studies underestimates the total value. The direct measurements for the obscured star-formation activity with the cold dust emission are thus required in all these there layers.

1.4 Scope of This Thesis

In this paper, we analyze the large dataset of multi-field deep ALMA Band 6/7 data including our original and archival data open for public by 2018 April. We obtain 1220 ALMA sources extracted from 1643 independent field maps in our analyses. With this largest ALMA dataset so far obtained, we investigate the Cold Universe in the scales of the ISM, the CGM, and the cosmic structure to unveil the total cosmic SFR density.

The structure of this paper is as follows. In Chap. 2, we describe the observations and data reduction. Chapter 3 shows the statistical size measurement results for the dust emission in the ISM scale. The detailed morphology in the ISM scale is analyzed with the high-resolution ALMA data in Chap. 4. In Chap. 5, we examine the emission from the dust and metal-enriched gas in the CGM scale. Chapter 6 outlines our statistical results of the number density and the clustering in the cosmic structure scale. In Chap. 7, we discuss the total cosmic SFR density based on our results obtained in the scales of the ISM, the CGM, and the cosmic structure. A conclusion composed of the summary is presented in Chap. 8.

Throughout this paper, we assume a flat universe with $\Omega_{\rm m} = 0.3$, $\Omega_\Lambda = 0.7$, $\sigma_8 = 0.8$, and $H_0 = 70$ km s^{-1} Mpc^{-1}. We use magnitudes in the AB system (Oke & Gunn 1983).

References

1. Puget J-L, Abergel A, Bernard J-P et al (1996) A&A 308:L5
2. Fixsen DJ, Dwek E, Mather JC, Bennett CL, Shafer RA (1998) ApJ 508:123
3. Hauser MG, Arendt RG, Kelsall T et al (1998) ApJ 508:25
4. Hauser MG, Dwek E (2001) ARA&A 39:249
5. Dole H, Lagache G, Puget J-L et al (2006) A&A 451:417
6. Dekel A, Birnboim Y, Engel G et al (2009) Nature 457:451
7. Muratov AL, Kereš D, Faucher-Giguère C-A et al (2015) MNRAS 454:2691
8. Lilly SJ, Carollo CM, Pipino A, Renzini A, Peng Y (2013) ApJ 772:119
9. Schaye J, Crain RA, Bower RG et al (2015) MNRAS 446:521
10. Sérsic JL (1963) Boletin de la Asociacion Argentina de Astronomia La Plata Argentina 6:41
11. Sérsic, JL (1968) Atlas de Galaxias Australes
12. Shen S, Mo HJ, White SDM et al (2003) MNRAS 343:978
13. Ferguson HC, Dickinson M, Giavalisco M et al (2004) ApJ 600:L107
14. Hathi NP, Malhotra S, Rhoads JE (2008) ApJ 673:686
15. Oesch PA, Bouwens RJ, Carollo CM et al (2010) ApJ 725:L150
16. Ono Y, Ouchi M, Curtis-Lake E et al (2013) ApJ 777:155
17. van der Wel A, Franx M, van Dokkum PG et al (2014) ApJ 788:28
18. Shibuya T, Ouchi M, Harikane Y (2015) ApJS 219:15
19. Bowler RAA, Dunlop JS, McLure RJ, McLeod DJ (2017) MNRAS 466:3612
20. Kawamata R, Ishigaki M, Shimasaku K, Oguri M, Ouchi M (2015) ApJ 804:103
21. Kawamata R, Ishigaki M, Shimasaku K et al (2018) ApJ 855:4
22. Hodge JA, Swinbank AM, Simpson JM et al (2016) ApJ 833:103
23. Hodge JA, Swinbank AM, Simpson JM et al (2015) ApJ 810:133
24. Hodge JA, Swinbank AM, Simpson JM et al (2015) ApJ 799:81

25. Fujimoto S, Ouchi M, Shibuya T, Nagai H (2017) ApJ 850:1
26. Targett TA, Dunlop JS, McLure RJ et al (2011) MNRAS 412:295
27. Targett TA, Dunlop JS, Cirasuolo M et al (2013) MNRAS 432:2012
28. Chen C-C, Smail I, Swinbank AM et al (2015) ApJ 799:194
29. Rujopakarn W, Dunlop JS, Rieke GH et al (2016) ApJ 833:12
30. González-López J, Bauer FE, Romero-Cañizales C et al (2017) A&A 597:A41
31. Momose R, Ouchi M, Nakajima K et al (2014) MNRAS 442:110
32. Leclercq F, Bacon R, Wisotzki L et al (2017) A&A 608:A8
33. Sravan N, Faucher-Giguère C-A, van de Voort F et al (2016) MNRAS 463:120
34. Turner ML, Schaye J, Crain RA et al (2017) MNRAS 471:690
35. Neri R, Downes D, Cox P, Walter F (2014) A&A 562:A35
36. Narayanan D, Turk M, Feldmann R et al (2015) Nature 525:496
37. Zhang Z-Y, Ivison RJ, George RD et al (2018) MNRAS 481:59
38. Falgarone E, Zwaan MA, Godard B et al (2017) Nature 548:430
39. Ivison RJ, Papadopoulos PP, Smail I et al (2011) MNRAS 412:1913
40. George RD, Ivison RJ, Smail I et al (2014) MNRAS 442:1877
41. Maiolino R, Gallerani S, Neri R et al (2012) MNRAS 425:L66
42. Heckman TM, Alexandroff RM, Borthakur S, Overzier R, Leitherer C (2015) ApJ 809:147
43. Li M, Bryan GL, Ostriker JP (2017) ApJ 841:101
44. Watson D, Christensen L, Knudsen KK et al (2015) Nature 519:327
45. Maiolino R, Carniani S, Fontana A et al (2015) MNRAS 452:54
46. Capak PL, Carilli C, Jones G et al (2015) Nature 522:455
47. Pentericci L, Carniani S, Castellano M et al (2016) ApJ 829:L11
48. Knudsen KK, Richard J, Kneib J-P et al (2016) MNRAS 462:L6
49. Matthee J, Sobral D, Boone F et al (2017) ApJ 851:145
50. Carniani S, Maiolino R, Amorin R et al (2018) MNRAS 478:1170
51. Smit R, Bouwens RJ, Carniani S et al (2018) Nature 553:178
52. Bowler RAA, Bourne N, Dunlop JS, McLure RJ, McLeod DJ (2018) MNRAS 481:1631
53. Carniani S, Maiolino R, Pallottini A et al (2017) A&A 605:A42
54. Lagache G, Puget J-L, Dole H (2005) ARA&A 43:727
55. Eales S, Lilly S, Webb T et al (2000) AJ 120:2244
56. Smail I, Ivison RJ, Blain AW, Kneib J-P (2002) MNRAS 331:495
57. Coppin K, Chapin EL, Mortier AMJ et al (2006) MNRAS 372:1621
58. Knudsen KK, van der Werf PP, Kneib J-P (2008) MNRAS 384:1611
59. Weiß A, Kovács A, Coppin K et al (2009) ApJ 707:1201
60. Perera TA, Chapin EL, Austermann JE et al (2008) MNRAS 391:1227
61. Hatsukade B, Kohno K, Aretxaga I et al (2011) MNRAS 411:102
62. Scott KS, Wilson GW, Aretxaga I et al (2012) MNRAS 423:575
63. Hatsukade B, Ohta K, Seko A, Yabe K, Akiyama M (2013) ApJ 769:L27
64. Furusawa H, Kosugi G, Akiyama M et al (2008) ApJS 176:1
65. Ono Y, Ouchi M, Kurono Y, Momose R (2014) ApJ 795:5
66. Carniani S, Maiolino R, De Zotti G et al (2015) A&A 584:A78
67. Reddy N, Dickinson M, Elbaz D et al (2012) ApJ 744:154
68. Lee K-S, Alberts S, Atlee D et al (2012) ApJ 758:L31
69. Davies LJM, Bremer MN, Stanway ER, Lehnert MD (2013) MNRAS 433:2588
70. Decarli R, Smail I, Walter F et al (2014) ApJ 780:115
71. Webb TM, Eales S, Foucaud S et al (2003) ApJ 582:6
72. Blain AW, Chapman SC, Smail I, Ivison R (2004) ApJ 611:725
73. Williams CC, Giavalisco M, Porciani C et al (2011) ApJ 733:92
74. Hickox RC, Wardlow JL, Smail I et al (2012) MNRAS 421:284
75. Miller TB, Hayward CC, Chapman SC, Behroozi PS (2015) ArXiv e-prints, arXiv:1501.04105
76. Schiminovich D, Ilbert O, Arnouts S et al (2005) ApJ 619:L47
77. Reddy NA, Steidel CC (2009) ApJ 692:778
78. Reddy NA, Steidel CC (2015) ApJ 803:34
79. Bouwens RJ, Illingworth GD, Oesch PA et al (2014) ApJ 793:115
80. Meurer GR, Heckman TM, Calzetti D (1999) ApJ 521:64
81. Le Floc'h E, Papovich C, Dole H et al (2005) ApJ 632:169
82. Smolčić V, Schinnerer E, Zamorani G et al (2009) ApJ 690:610

Table 6.5 Multiple images at $z = 2.60$ with the ALMA counterpart

ID	ID	R.A.	Decl.	$S^{obs}_{1.3mm}$	$S^{corr}_{1.3mm}$	$S^{corr}_{1.2mm}$	g_{475}	r_{625}	i_{775}	z_{850}	μ
(L07)	(This work)	(J2000)	(J2000)	(mJy)	(mJy)	(mJy)	(mag$_{AB}$)				
(1)	(2)	(3)		(4)	(5)	(6)	(7)				(8)
5.1	…	13:11:29.064	−1:20:48.64	0.18 ± 0.06†	0.049 ± 0.021	0.067 ± 0.029	26.29	25.53	25.37	25.17	5.7
5.2	A1	13:11:29.224	−1:20:44.24	0.30 ± 0.06	0.015 ± 0.006	0.020 ± 0.008	26.61	25.78	25.58	25.36	22.8
5.3	…	13:11:34.120	−1:20:20.96	<0.28	<0.093	<0.13	26.61	25.85	25.85	25.60	3.0

(1) and (2) ID names given in Limousin et al. [61] (L07) and this work, respectively. (3) R.A. and Decl. in the optical data. (4) Peak flux density SExtractor measurement (Sect. 6.1.1) with the primary beam correction. If no significant 1.32 mm fluxes are detected at the optical source positions, three sigma upper limits are presented. (5) Observed flux density at 1.32 mm on the uv-taper image that are measured by 2D Gaussian-fitting routine of IMFIT in CASA with the primary beam correction. The magnification uncertainty (Sect. 6.1.5) is included in the error. (6) Source flux density at 1.2 mm with the flux scaling from 1.32 mm to 1.2 mm (Sect. 6.1). (7) Total magnitudes in the *HST*/ACS bands. (8) Magnification factor at the peak of the optical flux

†Marginal detection of the 3σ level that is below our source selection limit for the cluster data ($>3.9\sigma$; see Sect. 6.1.2)

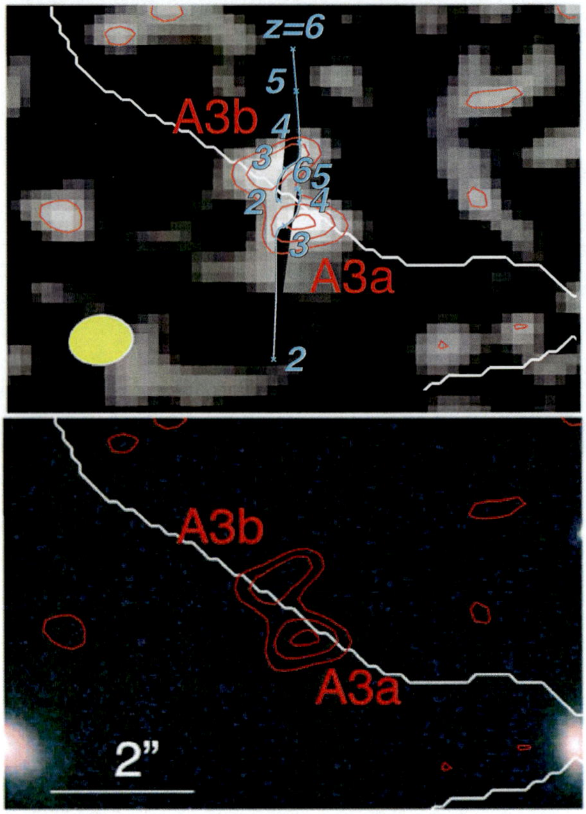

Fig. 6.10 Very strongly lensed source candidate, A3, found in A1689 (Fig. 11 of [9], reproduced by permission of AAS). The red contours show the ALMA 1.3 mm emission at the 2, 3, 4 σ levels. The white curves denote the critical lines for a $z = 3$ source that are estimated with our mass model. *Top*: ALMA continuum image of A3. The cyan tracks show the positions of multiple images for lensed sources at $z = 2, 3, 4, 5$, and 6 that are predicted with our mass model. The ALMA multiple image positions (red contours; A3a and A3b) are well reproduced by a lensed source at $z \sim 3$. *Bottom*: False-color image of HST optical data showing the same region as the top panel. The used images and color assignment are the same as Fig. 2.4. No optical counterparts of A3 are identified

Fig. 6.10. We find that the positions of the double peaks are reproduced very well by the multiple images at $z \sim 3$ with the best-fit redshift of $z = 3.07$. In the best-fit case, the magnification factor is very high, 159. We show the HST optical images in the bottom panel of Fig. 6.10, where we identify no optical counterparts of A3. In Table 6.6, we summarize the properties of A3.

Table 6.6 Very strongly lensed source candidate

ID	R.A	Decl.	$S^{obs}_{1.3mm}$	$S^{corr}_{1.3mm}$	$S^{corr}_{1.2mm}$	g_{475}	r_{625}	i_{775}	z_{850}	μ
	(J2000)	(J2000)	(mJy)	(mJy)	(mJy)	(mag_{AB})				
(1)	(2)		(3)	(4)	(5)	(6)				(7)
A3	13:11:26.897	−1:20:32.02	0.24±0.06	0.0015 ± 0.0007	0.0020 ± 0.0010	>28.56	>28.33	>28.24	>27.22	159.5

(1)–(7) correspond to (2)–(8) in Table 6.5. These values are for the A3a component (Fig. 6.10). Because A3a and A3b are confused in the uv-taper image, these fluxes are obtained on the image without uv-taper

6.2.5.3 Color and Luminosity Properties

To understand the population of the faint ALMA sources with the detectable optical-NIR continuum emission, we investigate color (i.e. SED) and luminosity properties of the optical-NIR counterparts, S1-15 and A1-2, First of all, we refer the public photo-z catalog of Williams et al. [62] estimated from the optical-NIR SEDs, and find that photo-z values of S11, S12, S15 are $1.54^{+0.08}_{-0.02}$, $1.57^{+0.07}_{-0.02}$, and $1.45^{+0.03}_{-0.02}$, respectively. We also find that photo-z values of S4 and S6 are $0.77^{+0.23}_{-0.03}$ and $0.67^{+0.01}_{-0.02}$, respectively, and thus that S4 and S6 reside at low redshift ($z < 1$). This indicates that at least $\sim$10% ($= 2/17$) optical-NIR counterparts are low-z objects. Nevertheless, because the majority of mm sources are thought to be high-z galaxies and AGNs [75], we interpret that the rest of 15 ($=17 - 2$) optical-NIR counterparts are candidates of high-z galaxies. We thus use high-z populations ($z > 1$) from the literature for the comparison of colors and luminosities. The major high-z populations are LAEs, BzKs, DRGs, SMGs, LBGs (including BX/BM), and AGNs. Note that LAEs, SMGs, and DRGs are generally very faint in optical continua. To perform a reliable comparison with our optical-NIR counterparts, we focus on the other populations of LBGs (BX/BM) and BzKs.

Comparison with the LBG BX/BM Populations
The BX and BM galaxies generally reside in the redshift ranges of $2.0 \lesssim z \lesssim 2.5$ and $1.5 \lesssim z \lesssim 2.0$, respectively [78], and overlapped with the LBG population, while the LBG population also includes sources at $z \gtrsim 3$. The BX and BM galaxies are defined by the color criteria of $G - R$ and $U_n - G$. We adopt these color criteria to identify the BM and BX galaxies in our optical-NIR counterparts, but the U_n, G and R bands are not included in the photometry bands of our data. Therefore, we alternatively adopt a photometry set of $BV - R_c i'$ and $U - BV$, following the procedure in Ly et al. [76]. The $R_c i'$ and BV photometry are defined by

$$R_c i' = -2.5\log\left[\frac{x_1 f_R + (1 - x_1 f_{i'})}{3630\ \mu Jy}\right],$$
$$BV = -2.5\log\left[\frac{x_2 f_B + (1 - x_2 f_V)}{3630\ \mu Jy}\right], \tag{6.11}$$

where f_X is the flux density per unit frequency (erg s^{-1} cm^{-2} Hz^{-1}) in a band 'X' corresponding to B, V, R_c, and i'. The coefficients of x_1 and x_2 are 0.207 and 0.314, respectively. Based on these photometric relations and the original color criteria in [78], we derive the color criteria of the BX objects as

$$\begin{aligned} BV - R_c i' &\leq 0.2(u - BV) + 0.4, \\ U - BV &\geq BV - R_c i' + 0.23, \\ BV - R_c i' &\geq -0.2, \\ U - BV &\leq BV - R_c i' + 1.0. \end{aligned} \tag{6.12}$$

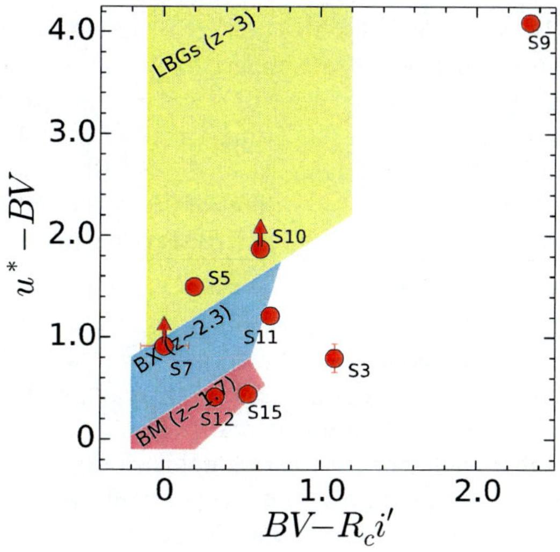

Fig. 6.11 Two-color diagram for our optical-NIR counterparts (Fig. 12 of [9], reproduced by permission of AAS). The cyan and magenta-shaded regions are the BX and BM selection windows that are defined in the $BV - R_c i'$ and $u^* - BV$ color plane [76]. The BX and BM galaxies reside at $z \simeq 2 - 2.5$ and $z \simeq 1.5 - 2.0$, respectively. The yellow shades indicate the approximate region of the selection window for $z \sim 3$ LBGs [77], where we assume that the original photometry of U_n, G and R bands are roughly the same as those of our u^*, B, and R_c bands, respectively

Similarly, we derive another color criteria of the BM objects as

$$\begin{aligned}
U \ -BV \ &\geq \ -0.1, \\
BV \ -R_c i' \ &\leq \ 0.70(U - BV) + 0.280, \\
BV \ -R_c i' \ &\leq \ -0.382(U - BV) + 0.853, \\
BV \ -R_c i' \ &\geq \ -0.2, \\
U \ -BV \ &\geq \ BV - R_c i' + 0.23.
\end{aligned} \tag{6.13}$$

We applying these alternative color criteria to our optical-NIR counterparts. There are 10 out of 17 optical-NIR counterparts that are observed with the photometric system of $u^* BV R_c i'$ required in the color criteria. Except for the obviously low-z objects of S4 and S6, we use 8 (=10 − 2) optical-NIR counterparts. In Fig. 6.11, we display the two color diagrams of these 8 sources together with the color selection criteria. We find that 6 sources fall in or near the selection windows of $z \sim 3$ LBGs or $z \sim 2$ BX/BM, where we include S11 at the border of the selection window. S9 locates at the upper right corner of Fig. 6.11. These colors around the S9 position are comparable to $z \gtrsim 4$ LBGs (see, e.g., [77]). It is thus likely S9 is explained by a LBG at $z \gtrsim 4$. On the other hand, S3 deviates from any selection windows. We conclude that 7 out of 8 optical-NIR counterparts have colors consistent with either $z \gtrsim 3$ LBGs or $z \sim 2$ BX/BM. Given the two low-z objects of S4 and S6,

70% (= 7/[8 + 2]) of the optical-NIR counterparts meet the color criteria for the LBG (BX/BM) population. We also compare magnitudes between the 10 optical-NIR counterparts and the known $z \gtrsim 3$ LBGs and $z \sim 2$ BX/BM. We find that the R_c magnitudes take the range of 23 − 27 mag in all of the 10 optical-NIR sources (Table 6.4), which is the typical brightness of $z \gtrsim 3$ LBGs and $z \sim 2$ BX/BM [77, 78]. We thus conclude that a majority (∼70%) of the optical-NIR counterparts of the faint ALMA sources belong to the LBG (BX/BM) population.

Comparison with the BzK Populations
In contrast with the LBG (BX/BM) population, the BzK population also includes moderately reddened dust-poor star-forming galaxies (a.k.a sBzK), in addition to blue star-forming galaxies, at $1.4 \lesssim z \lesssim 2.5$ [79]. We apply the color criteria of the BzK population to our optical-NIR counterparts. There are 10 out of the 17 optical-NIR counterparts that have the sufficient photometric measurements of B, z', and K bands (Table 6.4).[5] Removing obviously low-z objects of S4 and S6 again, we analyze 8 optical-NIR counterparts in this color selection.

Because the photometry system in SXDS used in our study is different from that of previous BzK studies (e.g., [79]), we again correct the color selection criteria for BzK galaxies. For this correction, we follow the same manner as [80] given by,

$$(B - z)_{\text{Daddi+04}} = (B - z)_{\text{SXDS}} + 0.3,$$
$$(z - K)_{\text{Daddi+04}} = (z - K)_{\text{SXDS}} + 0.1. \quad (6.14)$$

In Fig. 6.12, we present the $B - z$ and $z - K$ colors of the 8 optical-NIR counterparts. We find that 5 out of the 8 optical-NIR counterparts are classified as sBzK, while the rest of the three sources (S3, S9 and S10), have colors different from sBzK or pBzK galaxies whose redshift takes the range of $z \simeq 1.4 - 2.5$. Because S3 also shows the different colors from those of the LBGs and BX/BM populations (Sect. 6.2.5.3), S3 would fall at $z \lesssim 1.4$ that is out of the redshift window of BzK. Although the colors of S9 and S10 are similar to those of stars, in Sect. 6.2.5.3, the LBG and BX/BM galaxy selection results show that the colors of these two sources agree with $z \gtrsim 3$ galaxies. Thus, S9 and S10 would be galaxies at $z \gtrsim 3$ that fall out of the redshift window of BzK. In this way, we confirm that the results of the BzK selection are consistent with those of LBG and BX/BM galaxy selection. In summary, given the two low-z objects of S4 and S6, we find that 50%(=5/[8 + 2]) of the optical-NIR counterparts are classified as sBzK galaxies at $z \simeq 1.4 - 2.5$. We then again compare magnitudes of these optical-NIR counterparts (Table 6.4), and find that the optical and NIR magnitudes are consistent with those of the typical sBzK galaxies (e.g., [79]).

Comparison with the AGN Population
Although no X-ray nor radio continua are detected from our optical-NIR counterparts, we also study the mid-IR color property to further test for the AGN. This is because the bright mid-IR bright color is generally caused by the hot-dust emission, which

[5] The K band photometry is missed in the A1689 region misses.

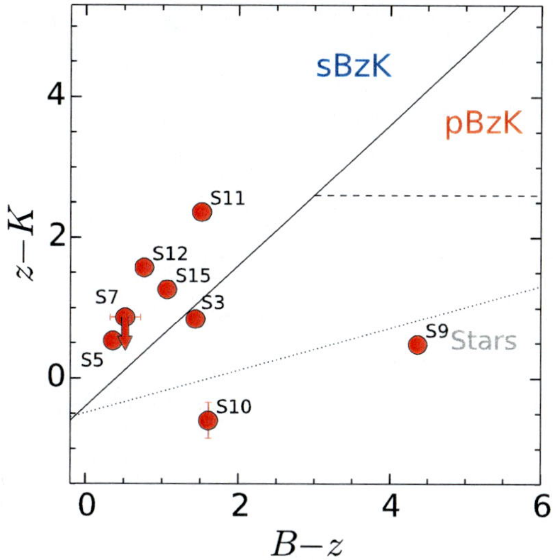

Fig. 6.12 Two-color diagram of the BzK galaxy selection in the SXDS photometry system (Fig. 13 of [9], reproduced by permission of AAS). The red circles denote the 7 optical-NIR counterparts. Star-forming galaxies at $z \simeq 1.4 - 2.5$, called sBzK galaxies, are selected in the region beyond the solid line defined by $(z - K) - (B - z) \geq -0.2$. The pBzK galaxies, passively evolving galaxies at $z \simeq 1.4 - 2.5$, are selected in the triangular region defined with the sold and dashed $(z - K > 2.5)$ lines at the upper right corner. Galactic stars fall in the region below the dotted line (See also [79, 80])

can be a good probe for the AGN. Six counterparts of S1, S6, S11, S12, S15, and A1 are detected in the mid IR of 8 and 24 μm bands, and we compare the mid-IR colors between the six mid-IR bright counterparts and the local templates of a starburst (Arp 220) a typical AGN (Mrk 231). In Fig. 6.13, we show the diagnostic color diagram of $S_{8.0\mu\mathrm{m}}/S_{4.5\mu\mathrm{m}}$ vs. $S_{24\mu\mathrm{m}}/S_{8.0\mu\mathrm{m}}$ that separates starburst (SB) and AGN populations [81]. The dashed and dotted lines are taken from Fig. 4 of [81], which indicates the color tracks of Mrk 231 and Arp 220 in the redshift range of 0.5−3.0. All of our six mid-IR bright counterparts fall near the color track of Arp 220 (SB), and far from that of Mrk 231 (AGN). This suggests that the six counterparts are not AGN, but SB.

IRX-β Relation

To evaluate the SED properties quantitatively, we calculate the rest-frame UV continuum slope β and the IR-to-UV luminosity ratio IRX($\equiv L_{\mathrm{IR}}/L_{1600}$) for 6 faint ALMA sources of S4, S6, S11, S12, S15, and A1 whose spectroscopic or photometric redshifts have been measured. Here we define L_{IR} as the integrated infrared flux density in a range of rest-frame 8–1000 μm. We estimate β with two broadband magnitudes, m_1 and m_2:

$$\beta = -\frac{m_1 - m_2}{2.5\log(\lambda_c^1/\lambda_c^2)} - 2, \tag{6.15}$$

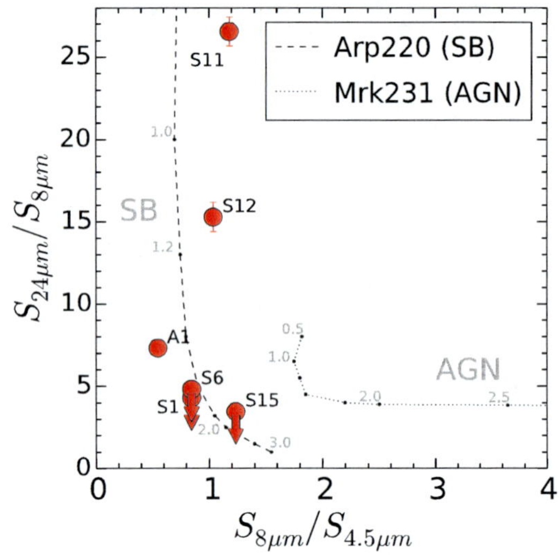

Fig. 6.13 Color-color diagram for the AGN diagnostics (Fig. 14 of [9], reproduced by permission of AAS). The red circles and upper limits are our five mid-IR bright counterparts. The dotted and dashed lines present the color tracks of Mrk 231 (AGN) and Arp 220 (SB), respectively, that are redshifted from $z \simeq 0.5$ to 3.0. The numbers labeled to the lines represent the redshifts

where λ_c^1 and λ_c^2 are the central wavelengths of the two broadband filters. We use $(m_1, m_2) = (u^*, B)$, (u^*, B), (V, i'), (V, i'), (V, i'), (r_{625}, z_{850}) for S4, S6, S11, S12, S15, and A1, respectively. These IRX and β values are shown in Fig. 6.14, together with those of optically selected star-forming galaxies at $z \sim 2$ and nearby (U)LIRGs. In Fig. 6.14, we also present 5.1, the multiple image of A1 (Table 6.5), whose ALMA continuum is also marginally detected ($\sim 3\sigma$). For the lensed sources of A1 and 5.1, we use the lensing magnification values estimated at the optical (submm) source positions, assuming that the positional offsets between the optical and submm sources are real, to derive the intrinsic optical (submm) luminosities.

In Fig. 6.14, all of the six faint ALMA sources show the IRX-β relations comparable to the Calzetti and SMC law curves. At a given β, all these six faint ALMA sources locates far below the (U)LIRGs. The clear difference between (U)LIRGs and the faint ALMA sources indicates that the faint ALMA sources are not the objects whose infrared luminosities are just fainter than (U)LIRGs by 1−2 order(s) of magnitude, i.e., miniature (U)LIRGs. We caution that these results hold on the 59% of the faint ALMA sources that have the optical-NIR counterparts. The rest of the 41% of the faint ALMA sources without optical-NIR counterparts could have low L_{1600} values, suggested by the fact of no optical-NIR counterparts, and IRX values as high as those of (U)LIRGs.

These 6 faint ALMA sources consist of 2 low-z sources (S4 and S6) and 4 high-z sources (S11, S12, S15, and A1), where we define low-z and high-z by $z < 1$ and $z \geq 1$, respectively. In Fig. 6.14, one low-z source of S4 is placed at the region of $z = 2$ optically selected star-forming galaxies [82], while the other low-z source of S6 has large IRX and β values. In contrast, all of the four high-z sources have the IRX-β relations well aligned with those of the $z = 2$ optically selected galaxies. These results imply that low-z sources would be widely distributed in the IRX-β plot, but that majority of high-z sources are similar to $z = 2$ optically selected galaxies.

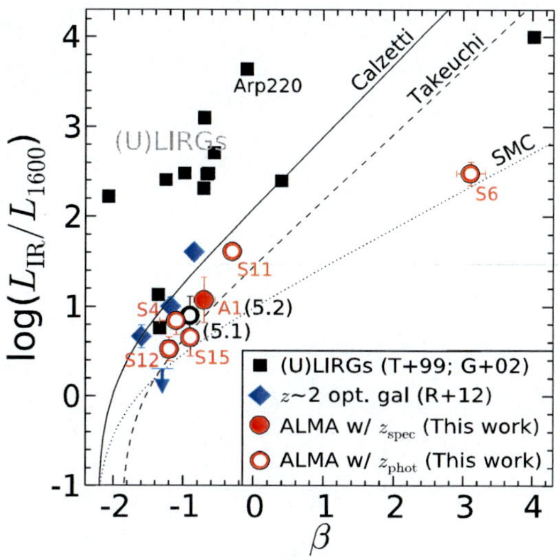

Fig. 6.14 Infrared-to-UV luminosity ratio IRX as a function of the UV slope β (Fig. 15 of [9], reproduced by permission of AAS). The red filled circle is the spectroscopically-confirmed A1 (5.2) source at $z_{\text{spec}} = 2.60$. The red open circles denote the faint ALMA sources with the photometric redshifts. The open black circle is the 5.1 source of the A1 multiple image system that is marginally detected in the ALMA data (Table 6.5). The blue diamonds (arrow) represent(s) the measurements (the upper limit) of the $z \sim 2$ optically-selected galaxies obtained by the stacking analysis [82]. The black squares indicate nearby (U)LIRGs [83, 84]. The solid, dashed, and dotted curves denote the IRX-β relation of the extinction curves for the Calzetti law [85], the updated Calzetti law [86], and the SMC law [87]. For the Calzetti law curve, we shift the original relation of the Calzetti law [85] by +0.24 dex, because [85] define their infrared luminosities at 40−120 μm that is different from our definition (8−1000 μm)

Note that we confirm that 5.2 (i.e., A1) and 5.1 are the multiple images of a single high-z source based on the similar IRX-β values.

What are the Optical-NIR Counterparts of the Faint ALMA Sources?
In Table 6.7, we summarize the results in this subsection. We find that 7 out of the 10 optical-NIR counterparts (including the low-z galaxies) meet the color criteria of LBG, BX/BM, and sBzK as well as the optical magnitudes consistent with these populations. Among the rest of the three counterparts, two are low-z objects and the remaining one is probably the foreground/background object due to the chance projection (Sect. 6.2.5.1). These results suggest that not all but the majority of the optical-NIR counterparts of our faint ALMA sources belong to the LBGs, BX/BM, and sBzKs populations. Moreover, neither AGN nor (U)LIRG signatures are identified from any optical-NIR counterparts.

We thus conclude that the majority of our faint ALMA sources with a detectable optical continuum (∼25 mag) are UV bright star-forming galaxies of LBG, BX/BM, and sBzKs with no AGN. The clustering analysis results in Sect. 6.2.4 show a weak clustering signal of $b_g < 3.5$ comparable to those of sBzKs and LBGs, which supports this conclusion.

Table 6.7 Properties of our optical-NIR counterparts

ID	R.A. (J2000)	Decl. (J2000)	λ_{obs} (mm)	$S^{obs}_{\lambda_{obs}}$ (mJy)	$S^{corr}_{1.2mm}$ (mJy)	z_{photo}	z_{spec}	sBzK	pBzK	BX/BM +LBG	AGN
			(1)	(2)	(3)	(4)	(5)	(6)	(7)	(8)	(9)
S1	34.612686	−4.583063	1.22	0.36 ± 0.07	0.43 ± 0.09	...	...	...	...	...	N
S2	34.613739	−4.588505	1.22	0.31 ± 0.09	0.31 ± 0.10	...	...	...	...	...	...
S3	34.437805	−5.494339	1.03	0.056 ± 0.015	0.032 ± 0.010	...	...	N	N	N	...
S4	34.446648	−5.007715	1.26	0.047 ± 0.014	0.044 ± 0.016	$0.77^{+0.23}_{-0.03}$	...	...	...	...	...
S5	34.446007	−5.008159	1.26	0.049 ± 0.014	0.050 ± 0.017	...	...	Y	N	Y	...
S6	34.387825	−5.220457	1.25	0.41 ± 0.11	0.42 ± 0.13	$0.67^{+0.01}_{-0.02}$	...	...	...	...	N
S7	34.411190	−4.745051	1.23	0.30 ± 0.08	0.14 ± 0.09†	...	...	Y′	N	Y′	...
S8	34.305553	−5.067293	1.30	0.29 ± 0.07	0.22 ± 0.09	...	...	...	...	...	...
S9	34.194702	−5.056259	1.25	0.43 ± 0.12	0.39 ± 0.15	...	...	N	N	Y′	...
S10	34.197121	−5.059778	1.25	0.31 ± 0.08	0.37 ± 0.10	...	...	N	N	Y′	...
S11	34.442795	−4.911066	1.23	0.53 ± 0.10	0.56 ± 0.11	$1.54^{+0.08}_{-0.02}$	...	Y	N	Y′	N
S12	34.441357	−4.912045	1.23	0.32 ± 0.09	0.29 ± 0.09	$1.57^{+0.07}_{-0.02}$	...	Y	N	Y	N
S13	34.747295	−4.858103	1.25	0.37 ± 0.10	0.26 ± 0.11	...	...	...	...	...	...
S14	34.274231	−4.860201	1.30	0.32 ± 0.09	0.43 ± 0.12	...	...	...	...	...	...
S15	34.587593	−5.318766	1.25	0.32 ± 0.09	0.21 ± 0.10	$1.45^{+0.03}_{-0.02}$	...	Y	N	Y	N
A1	197.871590	−1.345783	1.32	0.30 ± 0.06	0.020 ± 0.008††	...	2.60	...	...	...	N
A2	197.869086	−1.337143	1.32	0.31 ± 0.08	0.077 ± 0.033††	...	...	...	...	...	...

Notes (1): Wavelength in the observed frame. (2): Peak flux density of the SExtractor measurement (Sect. 6.1.1) with the primary beam correction at the observed wavelength. (3): The best-estimate source flux density at 1.2 mm. The source flux is estimated with the 2D Gaussian-fitting routine of *imfit* in CASA (Sect. 6.1.4) with the primary beam correction and the flux scaling to 1.2 mm (Sect. 6.1). For A1 and A2 sources, the lensing magnification corrections are also applied. The lensing magnification factors are estimated at the ALMA flux peak positions of A1 and A2. (4): Photometric redshift estimated by Williams et al. [62]. (5): Spectroscopic redshift obtained by Limousin et al. [61]. (6), (7), (8), and (9): "Y" indicates the objects that meet the color selection criteria of sBzK, pBzK, BX/BM, and AGN populations, respectively. "N" represents the sources escaping from the color space of the selection criteria. "Y′" is presented for the sources having colors consistent with the populations

†The photometry of *imfit* may be biased, due to the systematic noise. The SExtractor peak flux measurement suggests 0.32 ± 0.08 mJy. †† The magnification uncertainty (Sect. 6.1.5) is also included in the error

We caution that the optical-NIR counterparts are identified in 59% of our faint ALMA sources. In other words, the above conclusion can be applied roughly about a half of the faint ALMA sources, and the rest half of the faint ALMA sources are still unclear about their origins. One possibility is that the rest half of the faint ALMA sources are the dust obscured star-forming galaxies that have been missed in the optical–NIR observations. In fact, recent ALMA studies also report the existence of the dark objects in the optical–NIR wavelengths that are not detected in any HST bands even in CANDELS fields, but reliably detected with the ALMA (i.e. HST–dark objects; [88]). We discuss the existence of the objects missed in the optical–NIR observations in Sect. 7.4.

References

1. Kovács A, Chapman SC, Dowell CD et al (2006) ApJ 650:592
2. Coppin K, Halpern M, Scott D et al (2008) MNRAS 384:1597
3. Chapin EL, Pope A, Scott D et al (2009) MNRAS 398:1793
4. Planck Collaboration, Abergel A, Ade PAR et al (2011) A&A 536:A21
5. Chapman SC, Blain AW, Smail I, Ivison RJ (2005) ApJ 622:772
6. Yun MS, Scott KS, Guo Y et al (2012) MNRAS 420:957
7. Simpson JM, Swinbank AM, Smail I et al (2014) ApJ 788:125
8. Bertin E, Arnouts S (1996) A&A 117:393
9. Fujimoto S, Ouchi M, Ono Y et al (2016) ApJS 222:1
10. Hatsukade B, Ohta K, Seko A, Yabe K, Akiyama M (2013) ApJ 769:L27
11. Ono Y, Ouchi M, Kurono Y, Momose R (2014) ApJ 795:5
12. Carniani S, Maiolino R, De Zotti G et al (2015) A&A 584:A78
13. Watson D, Christensen L, Knudsen KK et al (2015) Nature 519:327
14. Austermann JE, Aretxaga I, Hughes DH et al (2009) MNRAS 393:1573
15. Austermann JE, Dunlop JS, Perera TA et al (2010) MNRAS 401:160
16. Ikarashi S, Ivison RJ, Caputi KI et al (2014) arXiv:1411.5038
17. Ikarashi S, Ivison RJ, Caputi KI et al (2015) ApJ 799:81
18. Oguri M (2010) PASJ 62:1017
19. Ishigaki M, Kawamata R, Ouchi M et al (2015) ApJ 799:12
20. Kochanek CS (1991) ApJ 373:354
21. Coe D, Benítez N, Broadhurst T, Moustakas LA (2010) ApJ 723:1678
22. Diego JM, Broadhurst T, Benitez N et al (2015) MNRAS 446:683
23. Gehrels N (1986) ApJ 303:336
24. Scott KS, Wilson GW, Aretxaga I et al (2012) MNRAS 423:575
25. Karim A, Swinbank AM, Hodge JA et al (2013) MNRAS 432:2
26. Simpson JM, Smail I, Swinbank AM et al (2015) ApJ 807:128
27. Oteo I, Zwaan MA, Ivison RJ, Smail I, Biggs AD (2015) arXiv:1508.05099
28. Shimizu I, Yoshida N, Okamoto T (2012) MNRAS 427:2866
29. Makiya R, Enoki M, Ishiyama T et al (2015) arXiv:1508.07215
30. Knudsen KK, van der Werf PP, Kneib J-P (2008) MNRAS 384:1611
31. Chen C-C, Cowie LL, Barger AJ et al (2013) ApJ 776:131
32. Scoville N, Aussel H, Brusa M et al (2007) ApJS 172:1
33. Hodge JA, Karim A, Smail I et al (2013) ApJ 768:91
34. Schechter P (1976) ApJ 203:297
35. Fixsen DJ, Dwek E, Mather JC, Bennett CL, Shafer RA (1998) ApJ 508:123
36. Kennicutt RC Jr (1998) ARA&A 36:189

37. Adelberger KL, Steidel CC, Giavalisco M et al (1998) ApJ 505:18
38. Robertson BE (2010) ApJ 716:L229
39. Mo HJ, White SDM (2002) MNRAS 336:112
40. Sheth RK, Tormen G (1999) MNRAS 308:119
41. Webb TM, Eales S, Foucaud S et al (2003) ApJ 582:6
42. Blain AW, Chapman SC, Smail I, Ivison R (2004) ApJ 611:725
43. Weiß A, Kovács A, Coppin K et al (2009) ApJ 707:1201
44. Williams CC, Giavalisco M, Porciani C et al (2011) ApJ 733:92
45. Hickox RC, Wardlow JL, Smail I et al (2012) MNRAS 421:284
46. Grazian A, Fontana A, Moscardini L et al (2006) A&A 453:507
47. Hayashi M, Shimasaku K, Motohara K et al (2007) ApJ 660:72
48. Quadri R, van Dokkum P, Gawiser E et al (2007) ApJ 654:138
49. Blanc GA, Lira P, Barrientos LF et al (2008) ApJ 681:1099
50. Furusawa J, Sekiguchi K, Takata T et al (2011) ApJ 727:111
51. Lin L, Dickinson M, Jian H-Y et al (2012) ApJ 756:71
52. Ouchi M, Shimasaku K, Okamura S et al (2004) ApJ 611:685
53. Adelberger KL, Steidel CC, Pettini M et al (2005) ApJ 619:697
54. Overzier RA, Bouwens RJ, Illingworth GD, Franx M (2006) ApJ 648:L5
55. Lee K-S, Giavalisco M, Gnedin OY et al (2006) ApJ 642:63
56. Gawiser E, Francke H, Lai K et al (2007) ApJ 671:278
57. Ouchi M, Shimasaku K, Furusawa H et al (2010) ApJ 723:869
58. Fry JN (1996) ApJ 461:L65
59. Peebles PJE (1993) Principles of physical cosmology. Princeton University Press, Princeton
60. Scoville N, Sheth K, Aussel H et al (2015) arXiv:1505.02159
61. Limousin M, Richard J, Jullo E et al (2007) ApJ 668:643
62. Williams RJ, Quadri RF, Franx M, van Dokkum P, Labbé I (2009) ApJ 691:1879
63. Ueda Y, Watson MG, Stewart IM et al (2008) ApJS 179:124
64. Furusawa H, Kosugi G, Akiyama M et al (2008) ApJS 176:1
65. Oliver SJ, Bock J, Altieri B et al (2012) MNRAS 424:1614
66. Simpson C, Martínez-Sansigre A, Rawlings S et al (2006) MNRAS 372:741
67. Ashby MLN, Stanford SA, Brodwin M et al (2013) ApJS 209:22
68. Bouwens RJ, Illingworth GD, Franx M et al (2009) ApJ 705:936
69. Bouwens RJ, Illingworth GD, Franx M et al (2015) ApJ 803:34
70. Downes AJB, Peacock JA, Savage A, Carrie DR (1986) MNRAS 218:31
71. Chen C-C, Cowie LL, Barger AJ, Wang W-H, Williams JP (2014) ApJ 789:12
72. Chen C-C, Smail I, Swinbank AM et al (2015) ApJ 799:194
73. Kneib J-P, van der Werf PP, Kraiberg Knudsen K et al (2004) MNRAS 349:1211
74. Swinbank AM, Smail I, Longmore S et al (2010) Nature 464:733
75. Casey CM, Narayanan D, Cooray A (2014) Phys. Rep. 541:45
76. Ly C, Malkan MA, Hayashi M et al (2011) ApJ 735:91
77. Steidel CC, Adelberger KL, Shapley AE et al (2003) ApJ 592:728
78. Steidel CC, Shapley AE, Pettini M et al (2004) ApJ 604:534
79. Daddi E, Cimatti A, Renzini A et al (2004) ApJ 617:746
80. Yuma S, Ohta K, Yabe K (2012) ApJ 761:19
81. Ivison RJ, Greve TR, Serjeant S et al (2004) ApJS 154:124
82. Reddy N, Dickinson M, Elbaz D et al (2012) ApJ 744:154
83. Trentham N, Kormendy J, Sanders DB (1999) AJ 117:2152
84. Goldader JD, Meurer G, Heckman TM et al (2002) ApJ 568:651
85. Meurer GR, Heckman TM, Calzetti D (1999) ApJ 521:64
86. Takeuchi TT, Yuan F-T, Ikeyama A, Murata KL, Inoue AK (2012) ApJ 755:144
87. Bouchet P, Lequeux J, Maurice E, Prevot L, Prevot-Burnichon ML (1985) A&A 149:330
88. Franco M, Elbaz D, Béthermin M et al (2018) A&A 620:A152

Chapter 7
Discussion

7.1 Insterstellar Medium Scale

7.1.1 Do Mergers Trigger Dusty Starbursts?

7.1.1.1 Close Companion and Star-Formation Mode

In Sect. 3.2.4, we find that about a half of our ALMA sources have the dusty rest-frame FIR-emitting regions largely separated from those of the rest-frame UV and/or optical emission. This may suggest a possibility of the dusty starbursts triggered by the major mergers. For testing the possibility, we investigate the rest-frame UV and optical morphology and the star-formation mode of our ALMA sources. Given that the pre-selection bias in the rest-frame UV and optical bands exists in the archive data, we use the OC5S-mmT sources whose central targets in the initial ALMA observations are selected in the submm/mm bands (Sect. 3.2.1).

Firstly, the OC5S-mmT sources are cross-matched with the 3D-HST sources [1]. The high spatial resolution of the HST images is required to obtain homogeneous and reliable morphology results. We identify that 56 OC5S-mmT sources in the 3D-HST source catalog. In Fig. 7.1, we show the false-color HST images of these 56 OC5S-mmT sources.

Secondly, we classify the morphology of the 56 OC5S-mmT sources and test whether they are the major mergers or not. If the HST map shows major merger pairs of the OC5S-mmT source, we regard it as the major mergers. For the major merger pair identification, we perform the following three steps: (i) selecting the optical-NIR objects the spatial offset within a radius of $2''.5$ from the OC5S-mmT, (ii) choosing the optical-NIR objects with the $\Delta z_{\rm phot} < 1$ from the object selected in (i), and (iii) identifying the optical-NIR objects with $M_{\rm star}$ greater than 10% of that of the object selected in (ii). Note that the radius of $2''.5$ in (i) is equal to $\sim$20 kpc at $z = 2.5$, which is used in Le Fèvre et al. [3] to classify the major merger system. Even with a larger radius of $6''$ ($\sim$50 kpc at $z = 2.5$) in (i), we confirm that no additional OS5S-mmT sources are identified, which ensures that we have not missed

S. Fujimoto, *Demographics of the Cold Universe with ALMA*, Springer Theses,
https://doi.org/10.1007/978-981-16-4979-0_7

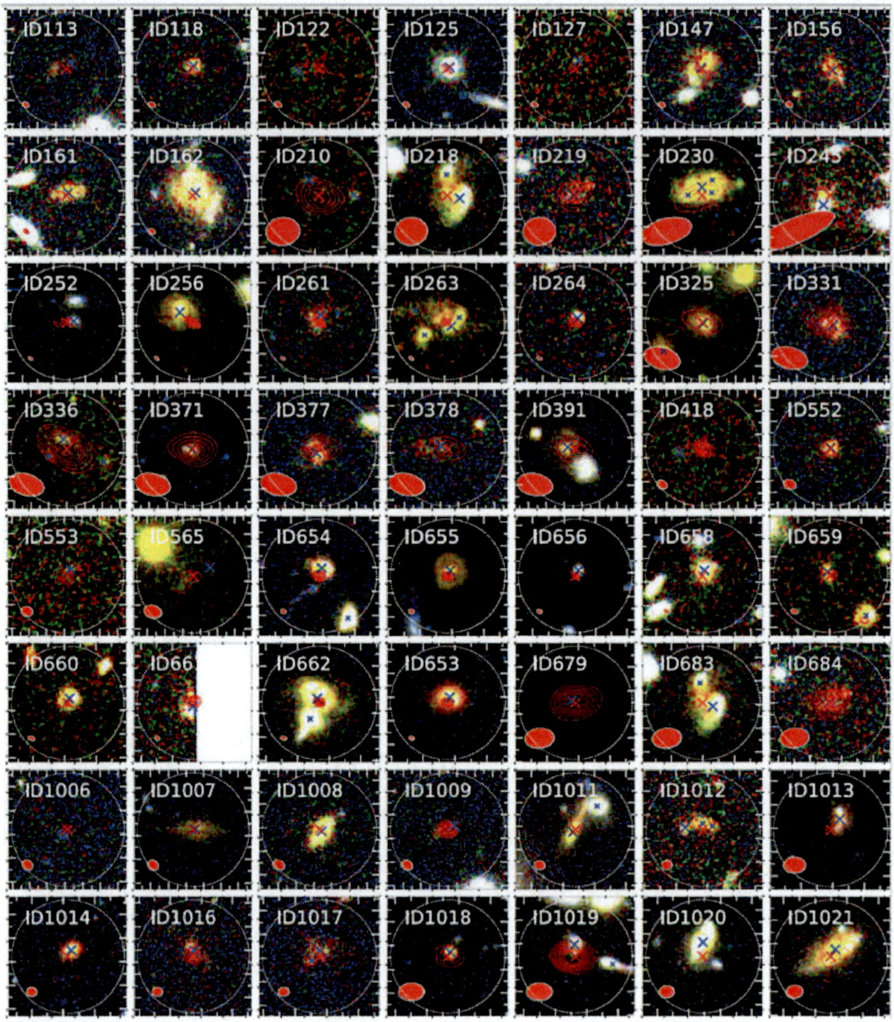

Fig. 7.1 $5'' \times 5''$ fake-color HST images of the 56 OC5S-mmT sources that are identified in the 3D-HST regions (red: H_{160}, green: J_{125}, blue: I_{814}). The red contours indicate ALMA mm band intensity from the 5 to 30σ levels with a 2σ-level step. The red cross represents the ALMA source center. The red ellipse shows the synthesized beam size. The white circle denotes the $2''.5$ search radius for the optical-NIR objects. The large blue cross presents the optical-NIR counterparts of the ALMA sources, while the small blue cross indicates the major merger pair. Note that ID661 is placed at the edge of the coverage of the HST observations. The figure is reproduced from Fig. 15 of Fujimoto et al. [2] by permission of AAS

any potential major mergers with our classification. We find that the major merger pairs exist in 27% (=15/56) of the OC5S-mmT sources. Because our classification of (i)–(iii) is limited by the spatial resolution of HST/H-band (0.″18) with which cannot identify close major merger pairs, we regard these 27% of the OC5S-mmT sources as "early/mid-stage major merger" We refer to the rest of 73% (=41/56) of the OC5S-mmT sources as "isolated galaxy".

Thirdly, we investigate the star-formation modes of the isolated galaxies and the early/mid-stage major mergers. In the left panel of Fig. 7.2, we show the $M_{\rm star}$ – SFR relation for the 56 OC5S-mmT sources. In the right panel, we present the histograms for the isolated galaxies and the early/mid-stage major mergers as a function of ΔMS, where we define ΔMS as the offset from the main sequence of the SFR–$M_{\rm star}$ relation with a unit of sSFR. We find that the early/mid-stage major mergers has a histogram similar to that of the isolated galaxies. The KS-test result for these two histograms shows that we cannot rule out the possibility that the same parent sample produces the samples of the early/mid-stage major mergers and the isolated galaxies.

A possible explanation for the the similar histograms is that the isolated galaxies are also dominated by the major mergers such as late-stage one. To further examine whether the morphology of the isolated galaxies show the hint of the mergers, we cross-match the isolated galaxies with the detail morphology catalog of Huertas-Company et al. [4] that evaluate the probabilities of the five classes of disk, spheroid, irregular, compact, and unclassifiable based on the Convolutional Neural Network technique. Here, we define the objects with the irregular probability over 50% as

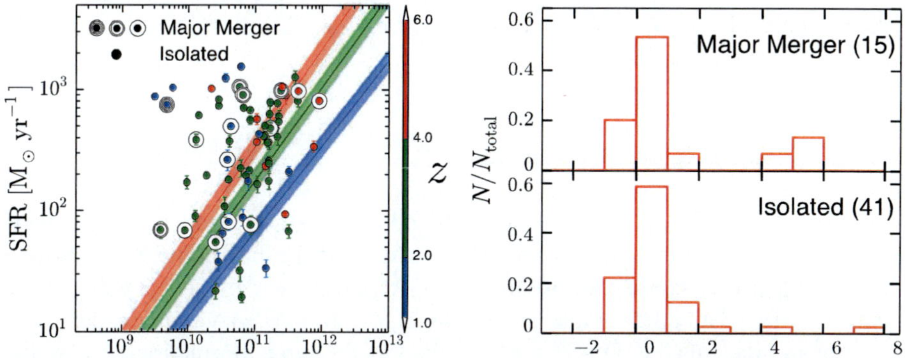

Fig. 7.2 *Left:* Our ALMA sources in the $M_{\rm star}$–SFR plane for investigating the star-formation mode (Fig. 16 of [2], reproduced by permission of AAS). The circles represent the 56 OC5S-mmT sources whose optical-NIR counterparts are detected in the 3D-HST regions. The number of the circles indicates the number of the multiple components in the major merger system that is indicated by the rest-frame UV-optical morphologies. The solid lines and the shade regions are the main sequences of the $M_{\rm star}$–SFR relation and the associated 1σ errors, respectively, at $z = 1$–2, 2–4, and 4–6 that are estimated by Speagle et al. [5]. The colors correspond to the redshift ranges in the color-bar scale. *Right:* Histograms of the isolated galaxies (bottom) and the major mergers (top) as a function of ΔMS, where ΔMS is the difference in sSFR from the main sequence

the merger-like morphology. The 25 isolate galaxies are studied in the morphology catalog, and we find that only 4 out of the 25 isolated galaxies are classified as the merger-like morphology. Even if we change the threshold of the irregular probability, we confirm that the fraction of the the merger-like morphology among the isolated galaxies is not changed significantly. This indicates that the isolated galaxies are not dominated by the late-stage mergers. If we combine our selections of the isolated galaxies with the merger-like morphology and the early/mid-stage major mergers, only $\sim$30–40% of our ALMA sources are classified as the major mergers. We thus conclude that not only major merger, but also other mechanism(s) trigger the dusty starburst.

We caution again that the spatial resolution of the data regulate the results of the visual-based morphology classification. Huertas-Company et al. [4] conduct the morphology classification based on the HST/H-band images with the spatial resolution of $0''.18$. On the other hand, recent ALMA observations achieve ultra-high resolutions of $0''.015 - 0''.05$ for high-z dust starbursts [6, 7]. Although the wavelengths are different between our analysis and these recent ALMA observations, we may obtain some hints for understanding the mechanisms of the dusty starburst. The ALMA ultra-high resolution maps unveil that four out of five high-z SMGs are resolved into several small clumps down to $\sim$200–300 pc scales with the high SFR density of $\sim$300–3000 $M_{\odot}\mathrm{yr}^{-1}\mathrm{kpc}^{-2}$ within the central kiloparcsec. Since such high SFR density values are comparable to those of the local major merger of Arp220, the main trigger can be explained by the major mergers [7]. Although the origins of the compact clumps identified in the ultra-high resolution maps are still under debated (cf. [8]), there recent ALMA results suggest the possibility that some of the isolated galaxies are also resolved into the merger-like morphologies after all by future ultra-high resolution observations.

7.1.1.2 Morphological Classification

We also check the L_{IR} dependence of the detail rest-frame optical morphological classification for our ALMA sources. To carry out the fair comparison based on the homogeneous dataset, we use the ASAGAO sources at $z = 1 - 3$ (Chap. 4). For the sake of statistically reliable results, we also verify ALMA sources identified in previous studies in GOODS-S [9, 10], and apply the same morphological classification to these ALMA sources as described in Sect. 4.1.3. The entire ALMA sample combining the ASAGAO and the previous ALMA sources span the L_{IR} range of $\sim 10^{11} - 10^{13}\ L_{\odot}$, where we divide the whole sample into two sub-samples: the high-z ULIRG ($L_{\mathrm{IR}} > 10^{12} L_{\odot}$) and LIRG ($L_{\mathrm{IR}} \leq 10^{12} L_{\odot}$) samples.

In Fig. 7.3, we show the histograms of the rest-frame optical morphology for the (U)LIRG samples. We also present the same histogram for the optically-selected galaxies classified in Kartaltepe et al. [12]. We find that the LIRG sample have higher and low fractions of the disk and the irregular/merging galaxy, respectively, than the ULIRG sample. The lower fraction of the irregular/merging galaxy indicates that the dusty star formation in the LIRG sample is not mainly triggered by the on-going

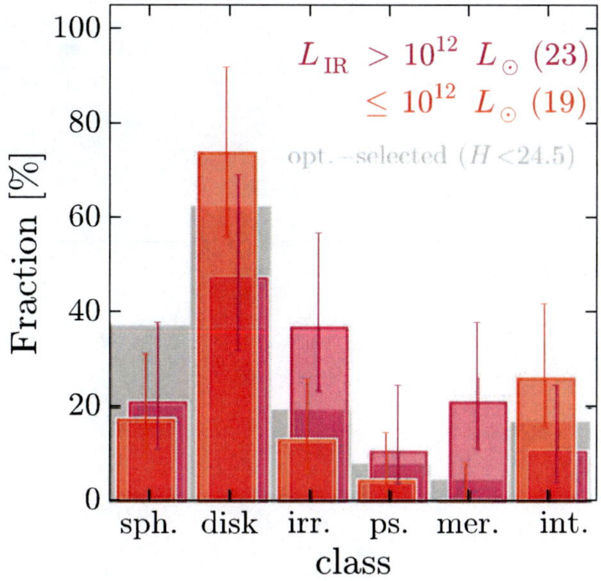

Fig. 7.3 Histograms of the rest-frame optical morphology with the deep HST/H-band images for the ALMA sources at $z = 1 - 3$ (Fig. 5 of [11], reproduced by permission of AAS). The morphological classification contains six classes: spheroid (sph.), disk, irregular (irr.), point source (ps.), merger (mer.), and interacting (int.) as in Kartaltepe et al. [12]. The magenta and red histograms indicate the high-z ULIRG ($L_{IR} > 10^{12}\ L_\odot$) and LIRG ($L_{IR} \leq 10^{12}\ L_\odot$) samples, respectively. The error-bars denote the Poisson uncertainty presented in Gehrels [13]. The number of galaxies in the ULIRG and LIRG samples are 23 and 19, respectively. The gray histogram is obtained from the entire sample of the >50,000 optically-selected galaxies ($H < 24.5$) in the morphology catalog of Kartaltepe et al. [12]. Note that the sum of the percentages exceeds 100% because the morphological classifications are not mutually exclusive. The magenta histogram is slightly shifted along the x-axis for clarity

merger processes. In fact, the fraction of disk galaxy in the LIRG sample is consistent with that of the optically-selected galaxies within the errors. These results imply that dusty star formation in the LIRG sample at high-z generally occurs in a secular star-formation mode in the regular disk galaxies. In contrast, the ULIRG sample has shows higher fractions of the irregular/merging galaxy than the other samples. This indicates that the on-going merger process plays a key role for triggering the dusty star formation in the ULIRG sample. Nonetheless, the disk galaxy is the most abundant population even in the ULIRG sample, implying that other mechanisms than the on-going merger process also cause the dusty star formation in the high-z ULIRGs. Note again that the rest-frame optical morphology classification is limited by the spatial resolution of HST/H-band ($\sim$0.″18). In both LIRG and ULIRG samples, the objects classified as the disk galaxy might be resolved into other morphology in future higher resolution observations.

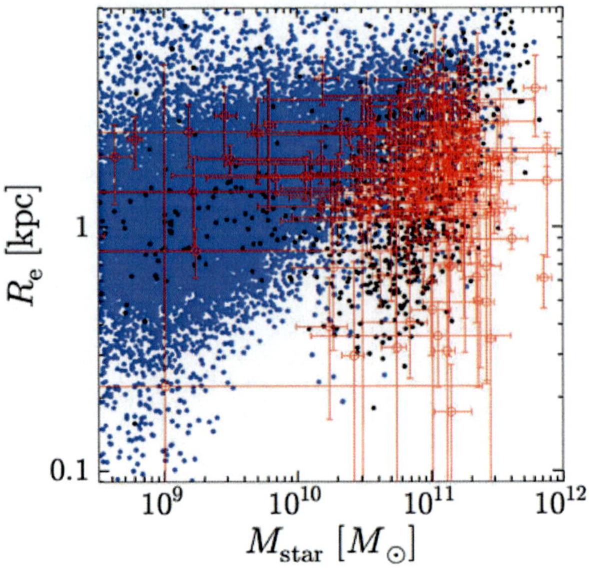

Fig. 7.4 Size−stellar mass relation for our ALMA sources and other galaxy populations (Fig. 13 of [2], reproduced by permission of AAS). The red open circles present the R_e values in the rest-frame FIR for the OC10S sources. The blue and black circles denote the R_e values in the rest-frame optical wavelength for the star-forming and quiescent galaxies at $z = 1-3$, respectively, in the 3D-HST regions obtained by Shibuya et al. [14]

7.1.2 Size and Stellar Mass Relation

In Sects. 3.2.5 and 4.2.3, we find that our ALMA sources have the rest-frame FIR size smaller than the rest-frame UV and optical sizes. To understand the role of the dusty star-forming galaxies in the context of the galaxy evolution, we examine the R_e as a function of M_{star} among the different wavelengths and galaxy populations.

Figure 7.4 presents the relation between R_e and M_{star}. For the dusty star-forming galaxies, we plot the OC10S sources with the R_e values in the rest-frame FIR wavelength in red circles. For comparison, we also plot the quiescent and the star-forming galaxies at $z = 1-3$ with the R_e values in the rest-frame optical wavelength in black and blue circles, respectively, that are obtained from the recent HST results [14]. The black and blue circles show the sequence of the quiescent and star-forming galaxies, where the latter is placed below the former on the R_e-M_{star} plane. We find that there is a good agreement between the majority of our ALMA sources and the sequence of the quiescent galaxies, albeit the wavelengths are different. Given that the abundant stars produced in the intense star-formation in the dusty starburst activities likely form the major part of the future stellar mass distribution in the host galaxies, it might be reasonable that our ALMA sources and the quiescent galaxies have the similar R_e values in the rest-frame FIR and optical wavelength that represent the current star-formation and the stellar mass distribution, respectively. This connection is also well aligned with the evolutionary scenario from the high-z dusty starbursts, the compact quiescent galaxies at $z \sim 2$, and to the local elliptical galaxies. (e.g., [15–22]). Interestingly, we also find some of our ALMA sources that fall between the sequences of the quiescent and the star-forming galaxies. We may be witnessing the transition phase from the star-forming to the quiescent galaxies by the rapid mass assembly via the dusty starburst.

7.1.3 Evolutions of Size and Morphology

In Sect. 7.1.2, we discuss the possibility of the direct connection from the dusty star-forming to compact quiescent galaxies. However, it is required to test whether the radial profiles of the total star formation can explain these size and morphological evolution among the high-z star-forming galaxies. In this Section, we derive the evolution track of the galaxy size and morphology in the rest-frame optical wavelength and discuss the evolutionary path from the high-z dusty star-forming galaxies to the compact quiescent galaxies.

For long years, it has been suggested an evolutionary path from high-z dusty star-forming galaxies, the compact quiescent galaxy (cQG) phase at redshift $z \sim 1-2$, and to local elliptical galaxies (e.g., [2, 8, 15–23]). In Sect. 4.2.2 we find that the dusty star-forming galaxies generally have the exponential-disk profile in the rest-frame optical wavelength of $R_{\rm e,opt} \sim 3$ kpc and $n_{\rm opt} \sim 1$, while the cQGs are generally have a different shape of the spheroidal profile of $R_{\rm e,opt} \lesssim 1$ kpc and $n_{\rm opt} \sim 4$ (e.g., [14, 24, 25]). To confirm the evolutionary path between the high-z dusty star-forming galaxies and the cQGs, we need to explain the increasing and decreasing trends of $n_{\rm opt}$ and $R_{\rm e,opt}$, respectively, at the same time.

Several evolution mechanisms have been discussed in the literature, such as "compaction" of the gas in star-forming galaxies due to disk instabilities [26], mergers of gas-rich galaxies (e.g., [17]), and inside-out growth of compact progenitors (e.g., [27, 28]). Here we alternatively test another evolutionary mechanism: the star-forming activity in the dusty star-forming galaxies alone regulates the evolutions of $n_{\rm opt}$ and $R_{\rm e,opt}$. We caution that high SFRs even up to an order of $\sim$1000 $M_\odot$/yr could arise in the dusty starburst systems, but the depletion time scale is generally as short as $\sim$100–200 Myr (e.g., [29]) in those intense star-formation activities. On the other hand, normal star-forming galaxies could keep a secure star-formation with a long depletion time of $\sim$1 Gyr (e.g., [30]). This indicates that the the majority of the stars may be produced by the secular star-formation, instead of the intense star-formation (e.g., [31–34]). We thus investigate the evolutions of $n_{\rm opt}$ and $R_{\rm e,opt}$ also for the high-z LIRGs ($L_{\rm IR} = 10^{11}-10^{12} L_\odot$), in addition to the high-z ULIRGs ($L_{\rm IR} > 10^{12} L_\odot$), which we regard as the galaxies under the secular star-formation and the intense modes, respectively.

To obtain the average properties of the high-z (U)LIRGs, we classify the 33 ASAGAO sources at $z = 1-3$ into two $L_{\rm IR}$ samples of $> 10^{12}$ $L_\odot$ and $L_{\rm IR} \leq 10^{12}$ that we regard as the representative samples of the high-z ULIRGs and LIRGs, respectively. For the average properties in the rest-frame optical and FIR wavelengths, we calculate the mean values of $M_{\rm star}$ and $L_{\rm IR}$ for the high-z ULIRG and LIRG samples. Following the average $L_{\rm IR}$ values, we evaluate the general relation between $L_{\rm IR}$ and Sérsic profiles in the rest-frame optical and FIR wavelengths from our best-fit results in Sect. 3.2. To evaluate the average property in the rest-frame UV wavelength, we use the deep HST/F606W image in GOODS-S and conduct the image-based stacking for the 33 ASAGAO sources. Because the massive galaxy centers are generally represented by the rest-frame FIR emission rather than the rest-frame UV emission (e.g., [22, 35]). we adjust the source centers for the stacking based on the ASAGAO source centers measured in ALMA Band 6. We then obtain the average values of the un-obscured rest-frame UV luminosity ($L_{\rm UV}$), effective radius ($R_{\rm e,UV}$), and Sérsic index ($n_{\rm UV}$) from the best-fit Sérsic profile of the stacked HST/F606W image with GALFIT. We also calculate the un-obscured (SFR$_{\rm UV}$) and the obscured SFR (SFR$_{\rm IR}$) from the average $L_{\rm UV}$ and $L_{\rm IR}$ with the equation in Straatman et al. [36] of

$$\mathrm{SFR_{UV}} = 2.4 \times 10^{-10}\, L_{\rm UV}(L_\odot), \tag{7.1}$$

$$\mathrm{SFR_{IR}} = 1.09 \times 10^{-10}\, L_{\rm IR}(L_\odot). \tag{7.2}$$

The left panel of Fig. 7.5 shows radial surface density profiles of $M_{\rm star}$ ($\Sigma M_{\rm star}$), SFR$_{\rm IR}$ (ΣSFR$_{\rm IR}$), and SFR$_{\rm UV}$ (ΣSFR$_{\rm UV}$) for the high-z LIRGs. We calculate the summation of ΣSFR$_{\rm UV}$ and ΣSFR$_{\rm IR}$ and also show the radial surface density profile of the total SFR (ΣSFR$_{\rm UV+FIR}$). The radial profile of $\Sigma M_{\rm star}$ is derived from the rest-frame optical Sérsic profile with HST/F160W-band (Sect. 4.2.2), based on the assumption that the source centers are the same in the rest-frame FIR and optical wavelengths. In these radial surface density profiles, we find in both high-z ULRIGs and LIRGs that the central star-forming region is dominated by ΣSFR$_{\rm IR}$ instead of ΣSFR$_{\rm UV}$.

Based on these radial profiles with assumptions of no radial motions of the stars and a constant SFR during a time scale Δt, we calculate the sum of ΣSFR$_{\rm UV+FIR} \times \Delta t$ and $\Sigma M_{\rm star}$ and evaluate the evolutions of $n_{\rm opt}$ and $R_{\rm e,opt}$ for the high-z ULRGs and LIRGs.

We assume two cases for Δt. The first case is the depletion time. From the gas fraction ($f_{\rm gas}$) as a function of $M_{\rm star}$ (e.g., [37, 38]), we estimate the gas mass ($M_{\rm gas}$) and calculate the depletion times to be ~210 and 640 Myr for the high-z ULIRGs and LIRGs, respectively. The second case is the cosmic time of ~1.0 Gyr from $z = 2$ to 1.5. This is because the star-formation could continue in a longer time than the depletion time owing to the gas supply from inflows. In fact, $f_{\rm gas}$ is suggested to be unchanged down to $z \sim 1$ by the recent bathtub model results, due to the the balance of the gas consumption by the star-formation and the gas inflow/outflow (e.g., [26]).

In the middle and right panels of Fig. 7.5, we show the results of the evolutions of $R_{\rm e,opt}$ and $n_{\rm opt}$ for the high-z LIRGs and ULIRGs in our model. For comparison, the distributions of $z \sim 1$–2 quiescent and $z \sim 2 - 3$ star-forming galaxies are also shown with the black and blue shaded regions, respectively. In the middle panel, we find that both high-z LIRGs and ULIRGs reproduce the decreasing trend of $R_{\rm e,opt}$ well. On the other hand, we find in the right panel that the increasing trend of $n_{\rm opt}$ is not reproduced. Since our model for the evolutions of $R_{\rm e,opt}$ and $n_{\rm opt}$ assumes no radial motions of the stars, these results suggest that forming the spheroidal profile

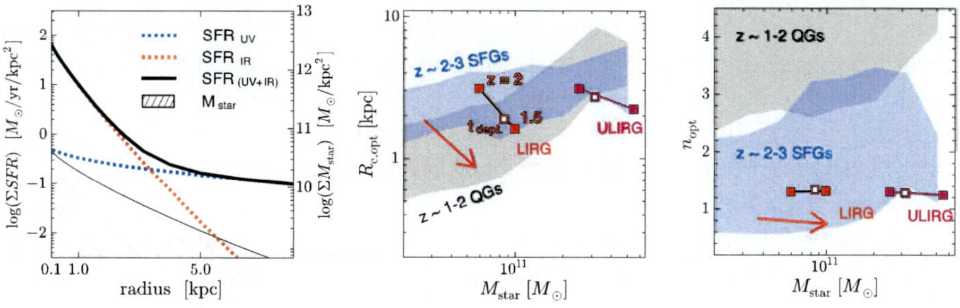

Fig. 7.5 ***(Left)*** Radial surface density profile of SFR and M_{star} for the high-z LIRGs. The red and blue dashed lines represent the radial profile of ΣSFR_{IR} from our best estimates in the rest-frame FIR wavelength (Sect. 4.2.1) and ΣSFR_{UV} from the stacking result with the deep HST/F606W image (see text), respectively. The black line indicates ΣSFR_{UV+FIR} that is obtained by combining the red and blue dashed lines. The black shade denotes ΣM_{star} estimated from our best estimates in the rest-frame optical wavelength (Sect. 4.2.2). ***(Middle)*** $R_{e,opt}$ evolution of the high-z (U)LIRGs. The red and magenta filled squares show the $R_{e,opt}$ evolutions of high-z ULIRG and LIRGs from $z = 2$ to $z = 1.5$ under the constant star-formation. The red and magenta open squares present the same $R_{e,opt}$ evolutions, but under the constant star-formation until the depletion time scales. The red and magenta lines indicate the evolutional tracks of ULIRG and LIRG, respectively. The blue and black shaded regions are 16th–84th percentiles of $z = 1.5-3.0$ star-forming (SFGs) and $z = 0.5-2$ quiescent galaxies (QGs) obtained from Shibuya et al. [14]. ***(Right)*** $n_{e(opt.)}$ evolution of high-z (U)LIRGs. The symbols and color assignments are the same as in the middle panel. The figure is reproduced from Fig. 6 of Fujimoto et al. [11] by permission of AAS

of $n_{opt} \sim 4$ requires other mechanisms such as dynamical dissipation from the star-forming to quiescent galaxies. In fact, a cQG recently discovered at $z = 2.1478$ turns out to be a rotationally supported, fast-spinning galaxy with $n_{opt} = 1.01^{+0.12}_{-0.06}$ [39]. Unless the angular momentum in a post star-forming galaxy that had an exponential-disk profile is lost by any dynamical dissipating events, it could be reasonable that the stellar distribution maintains the exponential-disk profile of $n_{opt} \sim 1$. Our model and the recent observational results consistently imply the existence of another step from the star-forming to quiescent galaxies to form the spheroidal stellar distribution. We caution that the initial M_{star} values in our model are massive relative to the main-sequence of the star-forming galaxies at $z \sim 2$. There remains a possibility that less massive, normal and/or dusty star-forming galaxies at higher redshifts are more likely the direct progenitor of the cQGs rather than $z \sim 2$ dusty star-forming galaxies used in our model. The future deep observations will examine the evolutions of size and morphology from the less massive, normal and/or dusty star-forming galaxies at higher redshifts to fully understand the origin of the cQGs.

7.1.4 Compact Dusty Bulge Versus AGN

In this Section, we discuss the origins of the compact component that we identify at the stacked source center for our ASAGAO sources (Sect. 4.1.5). In the right panel of Fig. 4.1, we find a slight offset between the best-fit and the stacked profiles near the center (radius $\lesssim 0.''1$). The offset likely follows the ALMA beam profile, indicating that this offset is explained by an unresolved compact component at the center. Interestingly, Hodge et al. [8] also find in the high-resolution ALMA maps that the spatial profiles in some of bright SMGs also agree with a combined profile of Gaussian + point source (PS).

The central compact component could be explained by the following two possible origins. One possibility is an AGN. In recent ALMA studies for local galaxies, a radius of ~150-pc scale dust-continuum emission has been observed from AGN and surrounding regions (e.g., [40]). In our ALMA HR map, the half-width-half-maximum (HWHM) of the ALMA beam is equal to ~800 pc at $z = 2$. If the spatial distribution of the dust continuum from the AGN and surrounding regions is not significantly changed between $z = 0$ and $z = 2$, the central AGN cannot be resolved in our ALMA HR map, and thus observed as a PS profile. The other possibility is a compact dusty bulge with a spheroidal ($n_{\rm FIR} \sim 4$) profile, probably related to the bulge formation at the center of the galaxy.

To understand which origin is more likely to explain the compact component at the center of the galaxy, we perform a two-component fitting for the $1.''6 \times 1.''6$ HR map of the stacked source by using GALFIT. We model two different profiles of PS+disk and bulge+disk. In both cases, the initial parameters for the disk component are taken from the best-fit values obtained from the single Sérsic profile fitting (Sect. 4.1.5). To obtain stable results, we fix $R_{\rm e}$ of the disk component and the central position of each component at the image center.

In the best-fit results, we find that the two different models produce almost the same minimum chi-square values. However, we find that $R_{\rm e}$ of the bulge component in the bulge+disk model reaches the minimum value in the fitting of $0.''0001$, indicating that the best-fit bulge+disk model can be regarded as the PS+disk profile. Based on the best-fit result with the PS+disk model, we estimate the contribution of the PS component to the total flux density to be $1.5 \pm 0.5\%$. Interestingly, recent studies 16 ALMA sources report a contribution of the AGN to the total flux density at the ~1-mm band to be ~0.5% from the SED decomposition of the star-formation and AGN [10], which is consistent with our above estimate within several factors. Moreover, Ueda et al. [41] report that a majority (~70%) of ALMA mm-selected galaxies in the $L_{\rm IR}$ range of $10^{11.5}$–$10^{12.8}$ $L_\odot$ are classified as X-ray AGNs. These results suggest that the compact component is likely explained by the central AGN. Although we cannot rule out a possibility that a PS-like very compact dusty bulge is formed at the center, it will be concluded in the future deep and high-angular resolution observations with ALMA.

7.2 Circumgalactic Medium Scale

In Chap. 5, we discover the [C II] halo around the normal star-forming galaxies at $z \sim 6$. In the following subsections, we discuss the physical origin of the [C II] halo.

7.2.1 Possible Origins of [CII] Halo

In Sect. 5.2.1, we find that the [CII] line emission extends to the ~10-kpc scale surrounding around the normal star-forming galaxies at $z = 5 - 7$. We also find that the radial profile of the extended [CII] line is potentially related to the Lyα halo. In contrast to the previous reports of the 10-kpc scale carbon abundance around the dusty starburst galaxies at $z \sim 2$ (e.g., [42, 43]), our results indicates the discovery of the universal cold carbon gas halos around early normal galaxies.

To understand the physical origin of the [CII] halo potentially associated with the Lyα halo, it is required to investigate the [CII] halo from the following two aspects: what supports the [CII] (and Lyα) line emissivity in the CGM? How the carbon abundance become enriched in the early phase of the CGM?

First, we focus on the first question of the [CII] line emissivity. The observational and theoretical studies suggest following five possible scenarios: (A) Satellite galaxies; (B) Circumgalactic (CG) PDR; (C) Photoionization; (D) Cold streams; and (E) Outflow. We illustrate these five possible scenarios in Fig. 7.6.

The first scenario explains that the sum of satellite galaxies causes the extended morphology (Fig. 7.6A). If there are satellite galaxies around the central galaxies, the Lyα and [CII] emission lines from these satellite galaxies could be observed as an extended structure around the central galaxy with a limited spatial resolution. In this scenario, the spatial distribution of the satellite galaxies defines the extended halo size and explains both extended components of the Lyα and [CII] lines. It may be also possible that these satellite components are the tidal tails produced by the past merging events.

The second scenario is a PDR extended over CG scale (Fig. 7.6B). Massive stars produce the ionizing photons, $h\nu > 13.6$ eV, and form the HII region on the central galactic scale. On the other hand, far-ultraviolet (FUV) photons, 6 eV $< h\nu <$ 13.6 eV, penetrate the surrounding ISM deeper than the ionizing photons and make the PDR distribute more widely than the HII region. Then, the [CII] line emission could be detected in the CG, if the PDR extends over the CG scale. Besides, the resonance scattering by the neutral hydrogen in the surrounding ISM (e.g., [45]) makes the Lyα line emission similarly spatially extended.

The third scenario is photoionization (Fig. 7.6C). This scenario is similar to scenario (B), but the HII region and the surrounding PDR is much more extended due to an existence of strong ionizing sources and/or the ISM properties differ from scenario (B). In this case, the Lyα line emission is extended due to the fluorescence (e.g., [46]), instead of the resonance scattering in scenario (B). Although the carbon may be doubly ionized with less [CII] line emission near the ionizing source centers in the highly ionized ISM, such deficit of the [CII] line emission is consistent with the recent ALMA results that many star-forming galaxies at $z > 5$ do not always show clear [CII] line detections at the stellar-continuum positions (e.g., [47]).

The fourth scenario invokes cold streams (Fig. 7.6D). The intense star-formation in high-z galaxies is suggested to be fed by a cold and dense gas (~10^4 K), dubbed cold streams, in cosmological hydrodynamical simulations (e.g., [48]). The cold

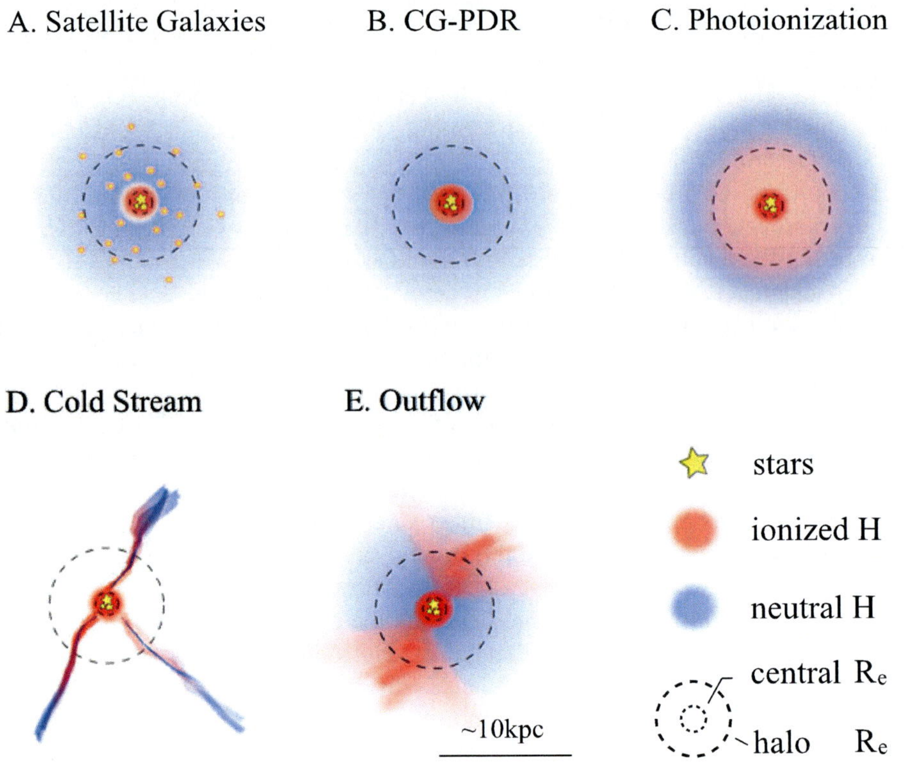

Fig. 7.6 Illustrations of five possible origins of the [C II] line emissivity in the CGM, with the potential association of the Lyα halo (Fig. 12 of [44], reproduced by permission of AAS). **A** satellite galaxies; **B** circumgalactic (CG) photodissociation region (PDR); **C** photoionization; **D** cold stream; and **E** outflow. The blue and red shades show the neutral and ionized hydrogen in ISM and CGM. The yellow stars represent the star-forming regions. The inner and outer dashed circle denote the effective radii (R_e) of the central and halo components of the [CII] line emission, respectively

streams radiate Lyα as well as [CII] emission lines powered by gravitational energy, and produce the extended Lyα and [CII] line structures around a galaxy. Moreover, the shock heating could be also caused by the cold stream, which also produces the Lyα and [CII] emission lines.

The last scenario is outflow (Fig. 7.6E). The ionized hydrogen and carbon powered by the star-formation and/or AGN feedback form the extended Lyα and [CII] emission lines, where a part of these line emission could be contributed by the associated process of the shock heating. We note that we cannot rule out the possibility that our ALMA sources have the past AGN activity and/or on-going faint AGNs, even though we choose ALMA sources not reported as AGNs.

7.2.2 From Observational and Theoretical Results

Hints from Observational Results:

In the observational results, we find that the [C II] line spatially extends more than both the rest-frame UV and FIR dust continuum up to a radius of at least ~7 kpc beyond the errors (Fig. 5.5). If we assume a constant line emissivity of [C II] at a given stellar continuum [49], a large gap appears between radial profiles of the stellar continuum and the [C II] line. This indicates that the large part of the [C II] line emissivity of the [C II] halo is insufficiently explained by the stellar continuum. Although the central to halo areas have different [C II] line emissivities at a given stellar continuum, such outer areas is expected to have a metallicity as low as ~1% of the galaxy center [50]. Even if low-mass, faint satellite galaxies produce the stellar continuum at the halo areas, the mass−metallicity relation [51] likewise suggests the lower metallicity in the satellite galaxies than in the central galaxy. Since the metallicity at the halo areas is expected to be significantly lower than the central areas [50, 51], Because less metallicity makes the [C II] line emissivity decreased at a given stellar continuum [52], the different [C II] line emissivities are furthermore difficult to explain the [C II] halo by the stellar continuum.

Moreover, our stacking results show that the $L_{[\mathrm{CII}]}/\mathrm{SFR}_{\mathrm{total}}$ ratio becomes higher towards halo areas where the high ratios cannot be explained either by the local star-forming or dwarf galaxies (Fig. 5.7). In contrast, the dwarf galaxies show the decreasing $L_{[\mathrm{CII}]}/\mathrm{SFR}_{\mathrm{total}}$ ratio as a function of SFR. This would be because the low-mass (~dwarf) galaxies maintain the harder UV radiation field than the central, massive galaxies due to the fact that low mass galaxies are generally compact [14, 25]. Not only the local studies, recent high-z observation results also suggest that the low mass galaxies have the high ionizing radiation field [53]. In the high ionizing radiation, the carbon is doubly or triply ionized, where the [C II] line emissivity is decreased and, again, hard to explain the extended [C II] line component by the satellite galaxies. Our observational results thus rule out the possibility of the scenario (A), supporting the rest of four scenarios of (B)−(E).

Hints from Theoretical Results:

In the zoom-in simulation results, the large part of the [C II] line luminosity in the [C II] halo is not reproduced (Fig. 5.10). If the current assumptions in the simulations for the [C II] line emissivity calculations are correct, the existence of the extended [C II] line requires additional physical mechanism(s) to explain it.

There are two possible mechanisms that are not included in the calculation of the [CII] line emissivity in the simulations, but may contribute to the existence of the extended [C II] line. The first mechanism is the shock heating (e.g., [54]). Although the shock process is followed by the hydro-dynamical calculation in the zoom-in simulation, the computation of the ionized carbon abundance in CLOUDY does not include the shock heating. Since the shock heating is caused in the galaxy merging or inflow/outflow processes, the scenarios of (A), (D), and (E) are supported. The second mechanism is the past/on-going AGN activities. The strong ionizing source of the past/on-going AGN activities forms a large area of the HII region and the surrounding PDR. Moreover, the AGN feedback may cause the shock heating, which also contributes to the [CII] line emissivity. In this case, the scenarios of (C) and (E) are supported. Although we cannot rule out the possibility that modifying any parts of the assumptions in the zoom-in simulation recovers the extended [CII] line luminosity, these results may suggest a hint that the shock heating and/or AGN activities contributes to the [CII] line luminosity in the extended halo component.

Note that 8 out of 12 sources in the ALMA-HST sample are placed at $5 < z < 6$, where the CMB effect is less than the zoom-in simulation results at $z = 6.0 - 7.2$. Because the CMB effect significantly reduces the diffuse [CII] line emission (e.g., [55, 56]), the redshift range slightly higher than the ALMA-HST sample may cause the insufficient [CII] line luminosity in the zoom-in simulation results.

We summarize the possible scenarios for the physical origin of the [CII] line emissivity in the [CII] halo. In the observational results, we rule out the scenario (A). In the zoom-in simulation results, if the assumptions in the calculation for the [CII] line emissivity are correct, all of the scenarios except for (B) are possible. The possible scenarios are thus (C), (D), and (E) with the current best estimates of both observational and theoretical results.

7.2.3 What Made the Primordial CGM Metal-Enriched?

Although there remain several possible scenarios that can give rise to [CII] emission in the CGM, it is important that that outflows are required in all cases to enrich the carbon abundance of the primordial CGM gas around the early galaxies. There are two modes of outflows, hot-mode and cold-mode outflows. The hot-mode outflow is defined as the outflow of ionized (hot) hydrogen gas that is heated by massive star/AGN radiation and supernova (SN) explosions. The cold-mode outflow is the outflow of neutral (cold) hydrogen gas that is pushed by the radiative and kinetic pressures given by massive-stars, AGNs, and SNe. Here, the majority of [CII] line is emitted in neutral (cold) hydrogen gas clouds. Moreover, the cosmic time at $z \sim 5-7$ ($\sim$1 Gyr) is too short to produce the [CII] emitting cold halos from hot gas expelled by the hot-mode outflow via recombination. Our finding of the [CII] halo in the 18 normal star-forming galaxies implies that outflows are dominated by cold-mode outflows in the early star-forming galaxies.

Since we find the consistent radial surface brightness profiles between the Lyα and [CII] emission lines (Sect. 5.2.4), the Lyα and [CII] halos may have the same physical origin. Future deep observations of both Lyα and [CII] emission lines for individual high-z galaxies will enable us to comprehensively understand the CGM metal enrichment also with the theoretical simulations that includes the radiative transfers of these emission lines.

7.3 Cosmic Structure Scale

In this Section, we assess the dust obscured star-forming galaxies via comparisons with the results of our ALMA and the optical-NIR observations.

7.3.1 Comparison with the IR Luminosity Function $0 \leq z \leq 4$

We first examine the reliability of our number counts, comparing with the IR luminosity function (LF). Gruppioni et al. [57] derive 11 IR LFs at $0 \leq z \leq 4$ based on the sources selected in *Herschel* PACS observations at 70, 100, and 160 μm, and obtain $L_{\rm IR}$ measurements by IR SED template fittings with the PACS and SPIRE (250–500 μm) photometric data. We estimate $S_{1.2\,{\rm mm}}$ of these sources at $0 \leq z \leq 4$ from the $L_{\rm IR}$ measurements, assuming the modified black body SED with $\beta_{\rm d} = 1.8^{+0.2}_{-0.3}$. If we adopt a constant $T_{\rm d}$, it causes a large uncertainty in the SED estimate due to the wide range of the $L_{\rm IR}$ measurements. We use the $L_{\rm IR} - T_{\rm d}$ relation [58] to estimate $T_{\rm d}$ for the modified blackbody with,

$$T_{\rm d}\,[{\rm K}] = 1.06 \times (\log L_{\rm IR}\,[L_{\odot}] - 9.5)^2 + 27.0. \tag{7.3}$$

This $L_{\rm IR} - T_{\rm d}$ relation is obtained by our modeling with the data points of Fig. 3 in Hwang et al. [58]. Because the $L_{\rm IR}$ range of the data points is $L_{\rm IR} \geq 10^{9.5}\,L_{\odot}$, we adopt a constant value of $T_{\rm d} = 27.0$ at $L_{\rm IR} < 10^{9.5}\,L_{\odot}$. From the $S_{1.2\,{\rm mm}}$ measurements, we derive eleven number counts from the eleven IR LFs. We refer this process as IR LF analysis.

Figure 7.7 presents the 1.2-mm number counts estimated from the IR LF analysis with the color filled regions. We extrapolate the 1.2-mm number counts below the limiting luminosity of each IR LF, and these extrapolations are color translucent regions in Fig. 7.7. Because most of the 1.2-mm number counts are extrapolations below 0.1 mJy, we perform following analyses at $S_{1.2\,{\rm mm}} \geq 0.1$ mJy. There are eight 1.2-mm number counts that include the extrapolations at $S_{1.2\,{\rm mm}} \geq 0.1$ mJy. We investigate the uncertainties of the extrapolations in these eight 1.2-mm number counts. In Gruppioni et al. [57], the shapes of these eight IR LFs ($0.6 \leq z \leq 4.2$) are determined by fixing the faint-end power law slope, $\alpha_{\rm p}$, at the value of -1.2. This $\alpha_{\rm p}$ value of -1.2 is obtained in the IR LF measurement for the local IR LF at $0 \leq z < 0.3$. However, the value of $\alpha_{\rm p}$ should evolve with redshift. In fact, recent optical studies have shown that $\alpha_{\rm p}$ of the LFs for star-forming galaxies are significantly steeper at high-z than low-z. Bouwens et al. [59] report that the $\alpha_{\rm p}$ value of the LF for star-forming galaxies is $\alpha_{\rm p} \sim -1.8$ at $z = 4$. To evaluate the uncertainties of extrapolations in the maximum case, we fix $\alpha_{\rm p} = -1.8$, and derive the best-fit function with the data points of the eight IR LFs in Gruppioni et al. [57]. From IR LFs to the 1.2-mm number counts for all of the IR LFs, we also estimate the uncertainty

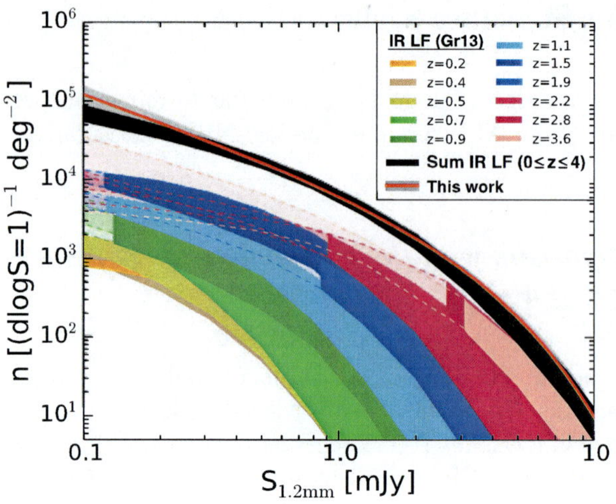

Fig. 7.7 Number counts at 1.2 mm estimated from the 11 IR LFs at $0 \le z \le 4$ (Gr13; [57]). The color regions show the eleven 1.2-mm number counts. The translucent regions with dashed lines denote the extrapolations for the eight 1.2-mm number counts that have the limiting luminosities at $S_{1.2\,\mathrm{mm}} > 0.1$ mJy. The black shade represents a total of these eleven 1.2-mm number counts, and the width of the black shade indicates the uncertainties. The red line with the gray shade presents our number counts at 1.2 mm with ALMA and the 1σ error obtained in Sect. 6.2.1

of the β_d of the modified black body. The color filled and translucent regions consist of these uncertainties of α_p and the β_d. We do not include the dispersion of the $L_\mathrm{IR} - T_\mathrm{d}$ relation for the uncertainty estimate because we aim to obtain average values statistically.

The black shade in Fig. 7.7 indicates a total of 1.2-mm number counts that sum over these eleven 1.2-mm number counts. The black shade is composed of the uncertainties of the eleven 1.2-mm number counts. Within the errors, the black shade is generally consistent with the red line and the gray region that is our ALMA 1.2-mm number counts. We thus conclude that our ALMA 1.2-mm number counts are reliable.

Using our reliable ALMA and the eleven 1.2-mm number counts, we also estimate the redshift distribution, n(z), for the faint ALMA sources. Here we first calculate the integrated fluxes of the eleven 1.2-mm number counts, and then obtain eleven contribution fractions to the integrated flux of our ALMA 1.2-mm number counts. These eleven contribution fractions are divided in four redshift bins at $0 \le z \le 4$ with $\Delta z = 1$. These contribution fractions of the four redshift bins are our n(z) estimate. We refer this process as n(z) analysis.

We first perform the n(z) analysis for the mm sources down to 1.0 mJy to compare with the n(z) estimate of SMGs in the literatures. The blue filled histograms in the top panel of Fig. 7.8 present the result of the n(z) analysis with $S_{1.2\,\mathrm{mm}} > 1.0$ mJy. The total contribution fraction up to $z = 4$ is $\sim$90%. We assume that the remaining $\sim$10% is contributed from the mm sources at $z > 4$, and show this remaining fractions with blue open histograms divided in the bins at $4 < z \le 6$. We find that the general shape

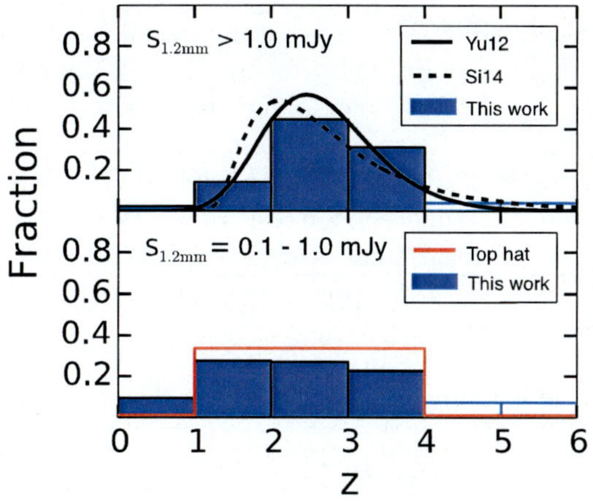

Fig. 7.8 Redshift distribution of SMGs ($S_{1.2\,\mathrm{mm}} \geq 1.0$ mJy; *Top panel*) and the faint ALMA sources ($0.1 \leq S_{1.2\,\mathrm{mm}} < 1.0$ mJy; *Bottom panel*) estimated from the n(z) analysis. The blue filled (open) histograms show the contribution fractions of of these mm sources at $0 \leq z \leq 4$ ($4 < z \leq 6$). The solid and dashed lines present the n(z) results of SMGs derived in Yun et al. [60] and Simpson et al. [61], respectively. The red open histogram indicates the top hat assumption of n(z) for the faint ALMA sources that we adopt through this paper

of our n(z) estimate is consistent with the previous n(z) studies for SMGs [60, 61]. We thus conclude that the n(z) analysis is reliable.

We next explore the n(z) analysis for the faint ALMA sources in the flux range of $0.1 - 1.0$ mJy, and show the result in the bottom panel of Fig. 7.8. We also illustrate the n(z) shape of the top hat assumption with the red histogram. This top hat assumption has the redshift range of $1 \leq z \leq 4$ with the median redshift value at $z_{\mathrm{med}} = 2.5$, which we adopt in the SFR estimate (Sect. 6.2.3) and the clustering analysis (Sect. 6.2.4). In the bottom panel of Fig. 7.8, the n(z) dispersion is larger than that of SMGs in the top panel, while the n(z) shape is generally consistent with the top hat assumption. We conclude that n(z) of the faint ALMA sources might have wider distribution than that of SMGs, and that our results of the analyses with the top hat assumption are reliable.

7.3.2 Comparison with the Stellar Mass Function $0 \leq z \leq 8$

Next, we investigate whether there are dust obscured star-forming galaxies missing in the optical-NIR observations. The dust obscured star-forming activity should be included in the total star formation rate, $\mathrm{SFR_{total}}$, given by

$$\mathrm{SFR_{total}} = \mathrm{SFR_{IR}} + \mathrm{SFR_{UV}}\ [M_{\odot}/\mathrm{yr}], \tag{7.4}$$

where $\mathrm{SFR_{IR}}$ ($\mathrm{SFR_{UV}}$) is the dust-obscured (UV) star formation rate.

We first estimate the $\mathrm{SFR_{total}}$ values of galaxies at $0 \leq z \leq 8$ from the M_{star} measurements, assuming the stellar mass functions (SMFs) and stellar mass (M_{star}) $-\mathrm{SFR_{total}}$ relation. For the SMF measurements, we use 7 SMFs of Muzzin et al. [62] that consist of star-forming and quiescent galaxies at $0 \leq z \leq 4$. We also use five SMFs at $4 \leq z \leq 8$ of Song et al. [63] who derive the SMFs from UV LFs and a median linear correlation between the rest-frame UV absolute magnitude at 1500 Å(M_{UV}) and M_{star} in the logarithmic scale. For the $M_{\mathrm{star}}-\mathrm{SFR}$ relation, we adopt a best-fit function form of star-formation main sequences (MSs) that are estimated by 25 independent studies [5],

$$\mathrm{SFR_{total}} = (0.84 - 0.026 \times t) \log M_{\mathrm{star}}\,[M_{\odot}] - (6.51 - 0.11 \times t), \quad (7.5)$$

where t is the cosmic age in Gyr.

We then calculate the $\mathrm{SFR_{IR}}$ values from the $\mathrm{SFR_{total}}$ estimates using the dust attenuation, A_{UV}, defined as

$$\begin{aligned} A_{\mathrm{UV}} &\equiv 2.5 \times \log(\mathrm{SFR_{IR}}/\mathrm{SFR_{UV}} + 1) \\ &= 2.5 \times \log(\mathrm{SFR_{IR}}/(\mathrm{SFR_{total}} - \mathrm{SFR_{IR}}) + 1). \end{aligned} \quad (7.6)$$

The A_{UV} values are estimated with the $M_{\mathrm{star}} - A_{\mathrm{UV}}$ relation of Pannella et al. [64],

$$A_{\mathrm{UV}}\,[\mathrm{mag}] = 1.6 \times \log M_{\mathrm{star}} - 13.5. \quad (7.7)$$

We convert the $\mathrm{SFR_{IR}}$ to L_{IR} values using the $\mathrm{SFR}-L_{\mathrm{IR}}$ relation [65], and then obtain the $S_{1.2\,\mathrm{mm}}$ values in the same manner as Sect. 7.7. From the $S_{1.2\,\mathrm{mm}}$ measurements, we derive twelve 1.2-mm number counts. We refer these process starting from the twelve SMFs as SMF analysis.

The color filled regions in Fig. 7.9 illustrates the seven 1.2-mm number counts estimated from the seven SMFs at $0 \leq z \leq 4$. The color translucent regions indicate extrapolations below the limiting M_{star} of each SMF. Since the M_{star} values are not directly measured in the five SMFs at $4 \leq z \leq 8$, the five 1.2-mm number counts estimated from these five SMFs are also presented with the color translucent regions. The width of the color filled and translucent regions are composed of the uncertainty of the extrapolations and β_{d} in the same manner as the previous section. There are three 1.2-mm number counts that include the extrapolations at $S_{1.2\,\mathrm{mm}} \geq 0.1$ mJy. These three 1.2-mm number counts are estimated from three SMFs at $2 \leq z \leq 4$. Because the α_{p} values of these three SMFs are poorly determined in Muzzin et al. [62], we fix the α_{p} value at -1.5 from the high-z studies [63]. We do not include the dispersion of the M_{star} - $\mathrm{SFR_{total}}$, M_{star} - A_{UV}, and SFR - L_{IR} relations, because here again our propose is to obtain average values statistically.

The black shade in Fig. 7.9 shows a total 1.2-mm number counts that sum over the twelve 1.2-mm number counts. The width of the black shade consists of the uncertainties from the twelve 1.2-mm number counts. In Fig. 7.9, we find that the total 1.2-mm number counts are consistent with our ALMA 1.2-mm number counts

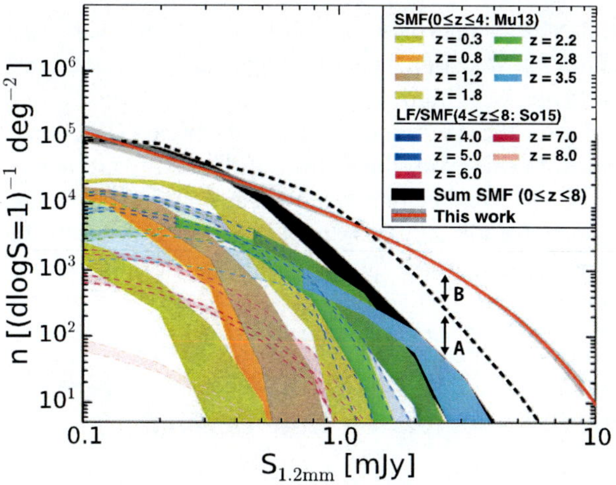

Fig. 7.9 Number counts at 1.2 mm estimated from the SMF analysis using the 12 SMFs at $0 \leq z \leq 8$ (Mu13; [62] and So15; [63]). The color filled regions show the seven 1.2-mm number counts from the SMFs of Mu13. The color translucent regions with the dashed lines denote the extrapolations for three out of the seven 1.2-mm number counts that have the limiting $M_{\rm star}$ above 0.1 mJy. The color translucent regions and the dashed lines also indicate the five 1.2-mm number counts from So15. The black shade represent the total of these twelve 1.2-mm number counts, and the width of the black shade indicates the uncertainties. The red line with the gray shade presents our ALMA 1.2-mm number counts and the 1σ errors obtained in Sect. 6.2.1. The two origins of the gap between the total and our ALMA 1.2-mm number counts are illustrated as the characters of "A" and "B" with arrows (see text for the details)

within the uncertainties at $S_{1.2\,\rm mm} \leq 0.4$ mJy. This result suggests that most of the faint ALMA sources at $S_{1.2\,\rm mm} \leq 0.4$ mJy can be explained by the optical star-forming galaxies if the extrapolations for the faint end of the present SMFs are correct in the reasonable range. On the other hand, Fig. 7.9 also shows that there is a large gap between the total and our ALMA 1.2-mm number counts at $S_{1.2\,\rm mm} > 0.4$ mJy beyond the uncertainties. There are two origins that cause this gap.

First origin is star-burst galaxies that are outliers on the $M_{\rm star}$−SFR relation with high $\rm SFR_{total}$ values. The higher $\rm SFR_{total}$ values at a fixed $M_{\rm star}$ produces higher $S_{1.2\,\rm mm}$ estimates than that of the normal star-forming galaxies on MS, which causes the 1.2-mm number counts shift to bright side in the SMF analysis. To evaluate this outlier effects in the maximum case, 50% of galaxies are applied with another $M_{\rm star}$−SFR relation that is scaled for SMGs (MS × 3; [66]). We then derive another total 1.2-mm number counts in the same manner as the SMF analysis. The black dashed line in Fig. 7.9 represents another total 1.2-mm number counts. We find that the gap still remains between another total and our ALMA 1.2-mm number counts even in the maximum case of the outlier effects. This result indicates that there exists another contributor of this gap which is the second origin.

The second origin is dust obscured star-forming galaxies that have been missed in the optical-NIR observations. In fact, recent studies report that ∼20% of the SMGs have no optical-NIR counterparts (e.g., [9, 67]). Moreover, we also confirm that ~40% of the faint ALMA sources have no optical-NIR counterparts in Sect. 6.2.5.1. Because the galaxies with high $M_{\rm star}$ values are highly dust obscured based on the $M_{\rm star} - A_{\rm UV}$ relation, these mm sources without optical-NIR counterparts might have high $M_{\rm star}$ values. In this case, the shapes of the SMFs are not well determined at the high $M_{\rm star}$ regime. We conclude that the gap at $S_{1.2\,\rm mm} > 0.4$ mJy consists of the two origins: (A) star-burst galaxies and (B) the uncertainty of the SMF shapes at the high $M_{\rm star}$ regime due to the dust obscured star-forming galaxies missing in the optical-NIR observations.

7.4 Cosmic Star-Formation Rate Density

The evolution of the cosmic star-formation rate density (SFRD) has been measured at high redshifts (e.g., [59, 68, 69]). These measurements are mainly based on the optical-NIR observations. However, there is a possibility that optical observations miss some of dust obscured star-forming galaxies. In this case, the SFRD measurements are underestimated. The direct observations for dust obscured star-formation are needed to measure the SFRD values correctly.

In Sect. 4, we find that the dusty star-formation occurs in a compact dusty disk embedded in a larger stellar disk. In Sect. 5, we discover the existence of the [C II] halo around $z \sim 6$ normal star-forming galaxies extended over the radius of ~10 kpc. We discuss the physical origin of these [C II] and dust continuum halos and conclude that these halos are not originated by the satellite galaxies. In Sect. 6.2.1, we derive the 1.2-mm number counts down to 0.02 mJy and almost fully resolve the CIB under the resolutions comparable to the ISM scale ($\simeq 0.''5$–$1.''0$). These results suggest that the dust obscured star-formation is taking place in the ISM scale, instead of widely distributed outside of the galaxies in the CGM and/or IGM scales, and we succeed in identifying the entire population of the ISM-scale dust obscured star-formation contributing to the CIB with ALMA. In this Section, we investigate the total cosmic star-formation rate density including the dust obscured part that may have been missed in the previous optical-NIR observations. Note that we reveal the possibility that there exist the dust obscured star-forming galaxies missing in the optical-NIR observations in the differential sense in Sect. 7.3.2, while we evaluate these dust obscured star-forming galaxies in the cumulative sense from the SFRD measurements in this Section.

Figure 7.10 shows the cosmic SFRD measurements estimated from previous optical-radio observations at $0 \leq z \leq 6$ [59, 70–73]. The black solid and dashed lines with gray shade present the best $\rm SFRD_{total}$ estimate of Hopkins and Beacom [68] who perform a functional model fitting to the previous SFRD measurements. To estimate the $\rm SFRD_{total}$ values including the dust obscuration, Hopkins and Beacom [68] calculate a sum of the $\rm SFRD_{UV}$ and $\rm SFRD_{IR}$ value that is obtained with the direct IR observations at $z \leq 1$. However, the $\rm SFRD_{total}$ measurements at $1 < z \leq 3$ are performed by adding the $\rm SFRD_{IR}$ value of $z = 1$ to the $\rm SFRD_{UV}$ values at $1 < z \leq 3$.

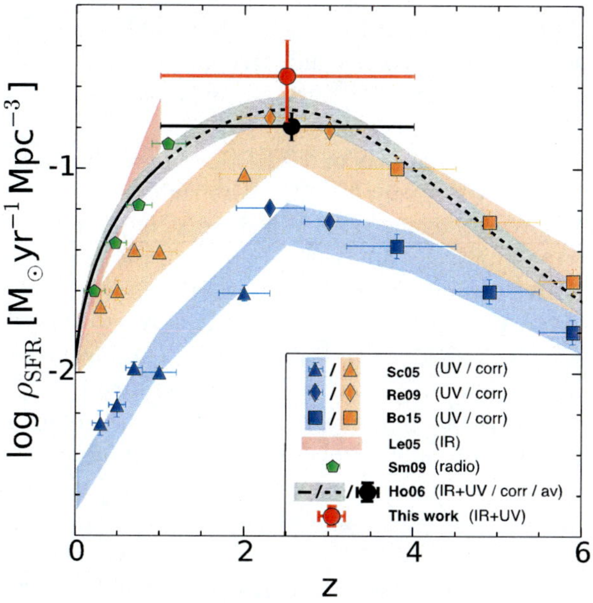

Fig. 7.10 Cosmic SFRD measurements from our ALMA 1.2-mm number counts and previous optical-radio observations. The red circle represents our SFRD$_{total}$ measurement at $1 \leq z \leq 4$ estimated from our SFRD$_{IR}$ and the average of the previous SFRD$_{UV}$ values [59, 70, 72]. The black solid and dashed lines denote the previous best estimate of the functional model fitting to the SFRD$_{total}$ and SFRD$_{corr}$ measurements at $0 \leq z \leq 6$, respectively [68]. The gray shade is the uncertainty of the previous best-estimate SFRD$_{total}$ value, and the black circle presents the average of the previous best-estimate SFRD$_{total}$ value at $1 \leq z \leq 4$. For presentation purposes, we slightly shift the black circle along the abscissa. The blue symbols are the SFRD$_{UV}$ measurements from the optical observations, while the orange symbols are dust corrected values estimated from the blue symbols. The blue and orange regions are the uncertainties of the SFRD$_{UV}$ and SFRD$_{corr}$ values that are presented in Fig. 18 of Bouwens et al. [59]. The light-red region shows the SFR$_{IR}$ measurement estimated from the direct IR observations at $z \leq 1$. The SFRD measurement from radio observations are also plotted in the green pentagons at $z \leq 1.3$

Moreover, the SFRD$_{total}$ measurements at $z > 3$ is estimated just by applying the dust corrections to the SFRD$_{UV}$ values (SFRD$_{corr}$). With these dust obscuration estimates at $z > 1$, the SFRD$_{total}$ values might miss the dust obscured star-formation.

We thus measure a SFRD$_{total}$ value from our ALMA 1.2-mm number counts, and compare the result with the value of Hopkins and Beacom [68]. In the SFRD$_{total}$ measurement, we first estimate the SFRD$_{IR}$ value from the integrated flux of our ALMA 1.2-mm number counts down to 0.02 mJy ($22.9^{+6.7}_{-5.6}$ Jy deg^{-2}; see Sect. 6.2.3). We convert the integrated flux to an integrated L_{IR} value with the assumption of $z = 2.5$, $\beta = 1.8$, and $T_d = 35$K. We then estimate the SFR$_{IR}$ value from the integrated L_{IR} value. From SFR$_{IR}$ to SFRD$_{IR}$, we adopt the top hat assumption at $1 \leq z \leq 4$ for the survey volume that is confirmed as reliable in the Sect. 7.3.1. We combine this SFRD$_{IR}$ value and a volume weighted average SFRD$_{UV}$ value at $1 \leq z \leq 4$ from the literatures in Bouwens et al. [59]. We then obtain the SFRD$_{total}$ value at

$1 \leq z \leq 4$, and plot the result in Fig. 7.10 with the red circle. A volume-weighted average SFRD$_{\rm total}$ value of Hopkins and Beacom [68] at $1 \leq z \leq 4$ is also plotted in Fig. 7.10 with the black circle to fairly compare our and the previous SFRD$_{\rm total}$ values.

In Fig. 7.10, we find the excess of our SFRD$_{\rm total}$ value to the average of the previous best-estimate SFRD$_{\rm total}$ value by 40^{+20}_{-40}%. Although there is no excess within 1σ error, this result provides us the same indication as the result in the SMF analysis that there might be the dust obscured star-forming galaxies missing in the optical-NIR observations. In fact, the light-red region in Fig. 7.10 from the direct IR observations shows the similar excess to the orange regain that presents the SFRD$_{\rm corr}$ values at $z \leq 1$. Moreover, the simple steps in this SFRD$_{\rm total}$ estimates include much less uncertainties than the uncertainties of the SMF analysis except for the uncertainty of the survey volume. To evaluate the uncertainty of the survey volume, we also calculate these SFRD$_{\rm total}$ values with another top hat assumption at $0 \leq z \leq 6$. We confirm that the excess still remains even with another top hat assumption because both of our and the average of the previous best-estimate SFRD values are systematically decreased according to the increase of the survey volume. We thus conclude that there exist the dust obscured star-forming galaxies missing in the optical-NIR observations with 1σ significance level.

The error bars of our SFRD$_{\rm total}$ value consist of the uncertainties from the average of the previous best-estimate SFRD$_{\rm UV}$ and our SFRD$_{\rm IR}$ measurements. The upper error of our SFRD$_{\rm IR}$ measurement is defined by the upper limit of the CIB value from the *COBE* observation. If the number count studies with ALMA reveal the flattening and/or truncation at $S_{\rm 1.2\,mm} \lesssim 0.02$ mJy (Sect. 6.2.3), we can set stronger upper limit than the present one. On the other hand, the lower error of our SFRD$_{\rm IR}$ measurement is estimated from the integrated flux error of our ALMA 1.2-mm number counts down to 0.02 mJy. Note that we confirm the reliability of our ALMA 1.2-mm number counts only at $S_{\rm 1.2\,mm} \geq 0.1$ mJy with the IR LF analysis in Sect. 7.3.1. We thus also include the difference of the integrated fluxes down to 0.02 mJy between our ALMA and the total 1.2-mm number counts as part of the lower error. This lower error and the center value of the SFRD$_{\rm IR}$ measurements can be increased as the ALMA observations provide us deeper data than $S_{\rm limit} = 0.02$ mJy. These upper and lower limit improvements with further ALMA observations will enable us to address the excess of the SFRD$_{\rm total}$ values beyond 1σ level.

References

1. Momcheva IG, Brammer GB, van Dokkum PG et al (2016) ApJS 225:27
2. Fujimoto S, Ouchi M, Shibuya T, Nagai H (2017) ApJ 850:1
3. Le Fèvre O, Abraham R, Lilly SJ et al (2000) MNRAS 311:565
4. Huertas-Company M, Gravet R, Cabrera-Vives G et al (2015) ApJS 221:8
5. Speagle JS, Steinhardt CL, Capak PL, Silverman JD (2014) ApJS 214:15
6. Speagle JS, Steinhardt CL, Capak PL, Silverman JD (2016) ApJ 822:36
7. Iono D, Yun MS, Aretxaga I et al (2016) ApJ 829:L10

8. Hodge JA, Swinbank AM, Simpson JM et al (2016) ApJ 833:103
9. Hodge JA, Karim A, Smail I et al (2013) ApJ 768:91
10. Dunlop JS, McLure RJ, Biggs AD et al (2017) MNRAS 466:861
11. Fujimoto S, Ouchi M, Kohno K et al (2018) ApJ 861:7
12. Kartaltepe JS, Mozena M, Kocevski D et al (2015) ApJS 221:11
13. Gehrels N (1986) ApJ 303:336
14. Shibuya T, Ouchi M, Harikane Y (2015) ApJS 219:15
15. Lilly SJ, Eales SA, Gear WKP et al (1999) ApJ 518:641
16. Genzel R, Baker AJ, Tacconi LJ et al (2003) ApJ 584:633
17. Tacconi LJ, Genzel R, Smail I et al (2008) ApJ 680:246
18. Hickox RC, Wardlow JL, Smail I et al (2012) MNRAS 421:284
19. Toft S, Smolčić V, Magnelli B et al (2014) ApJ 782:68
20. Chen C-C, Smail I, Swinbank AM et al (2015) ApJ 799:194
21. Chen C-C, Smail I, Swinbank AM et al (2015) ApJ 799:81
22. Barro G, Kriek M, Pérez-González PG et al (2016) ApJ 827:L32
23. Barro G, Kriek M, Pérez-González PG et al (2015) ApJ 810:133
24. van Dokkum PG, Franx M, Kriek M et al (2008) ApJ 677:L5
25. van der Wel A, Franx M, van Dokkum PG et al (2014) ApJ 788:28
26. Dekel A, Mandelker N (2014) MNRAS 444:2071
27. Nelson E, van Dokkum P, Franx M et al (2014) Nature 513:394
28. Wellons S, Torrey P, Ma C-P et al (2015) MNRAS 449:361
29. Carilli CL, Walter F (2013) ARA&A 51:105
30. Tacconi LJ, Genzel R, Neri R et al (2010) Nature 463:781
31. Elbaz D, Dickinson M, Hwang HS et al (2011) A&A 533:A119
32. Rujopakarn W, Rieke GH, Eisenstein DJ, Juneau S (2011) ApJ 726:93
33. Wuyts S, Förster Schreiber NM, van der Wel A et al (2011) ApJ 742:96
34. Schreiber C, Elbaz D, Pannella M et al (2016) A&A 589:A35
35. Tadaki K-I, Genzel R, Kodama T et al (2017) ApJ 834:135
36. Straatman CMS, Spitler LR, Quadri RF et al (2016) ApJ 830:51
37. Tacconi LJ, Neri R, Genzel R et al (2013) ApJ 768:74
38. Dessauges-Zavadsky M, Zamojski M, Schaerer D et al (2015) A&A 577:A50
39. Toft S, Zabl J, Richard J et al (2017) Nature 546:510
40. Izumi T, Kohno K, Aalto S et al (2015) ApJ 811:39
41. Ueda Y, Hatsukade B, Kohno K et al (2018) ApJ 853:24
42. Ivison RJ, Papadopoulos PP, Smail I et al (2011) MNRAS 412:1913
43. Falgarone E, Zwaan MA, Godard B et al (2017) Nature 548:430
44. Fujimoto S, Ouchi M, Ferrara A et al (2019) ApJ 887:107
45. Xue R, Lee K-S, Dey A et al (2017) ApJ 837:172
46. Cantalupo S, Porciani C, Lilly SJ, Miniati F (2005) ApJ 628:61
47. Carniani S, Maiolino R, Amorin R et al (2018) MNRAS 478:1170
48. Dekel A, Birnboim Y, Engel G et al (2009) Nature 457:451
49. De Looze I, Cormier D, Lebouteiller V et al (2014) A&A 568:A62
50. Pallottini A, Ferrara A, Bovino S et al (2017) MNRAS 471:4128
51. Mannucci F, Cresci G, Maiolino R, Marconi A, Gnerucci A (2010) MNRAS 408:2115
52. Vallini L, Gallerani S, Ferrara A, Pallottini A, Yue B (2015) ApJ 813:36
53. Stark DP, Richard J, Siana B et al (2014) MNRAS 445:3200
54. Appleton PN, Guillard P, Boulanger F et al (2013) ApJ 777:66
55. da Cunha E, Groves B, Walter F et al (2013) ApJ 766:13
56. Zhang Z-Y, Papadopoulos PP, Ivison RJ et al (2016) Royal Society Open Science 3:160025
57. Gruppioni C, Pozzi F, Rodighiero G et al (2013) MNRAS 432:23
58. Hwang HS, Elbaz D, Magdis G et al (2010) MNRAS 409:75
59. Hwang HS, Elbaz D, Magdis G et al (2015) ApJ 803:34
60. Yun MS, Scott KS, Guo Y et al (2012) MNRAS 420:957
61. Simpson JM, Swinbank AM, Smail I et al (2014) ApJ 788:125

62. Muzzin A, Marchesini D, Stefanon M et al (2013) ApJ 777:18
63. Song M, Finkelstein SL, Ashby MLN et al (2015) arXiv:1507.05636
64. Pannella M, Elbaz D, Daddi E et al (2015) ApJ 807:141
65. Kennicutt RC Jr (1998) ARA&A 36:189
66. da Cunha E, Walter F, Smail IR et al (2015) ApJ 806:110
67. Chen C-C, Smail I, Ivison RJ et al (2016) arXiv:1601.02630
68. Hopkins AM, Beacom JF (2006) ApJ 651:142
69. Madau P, Dickinson M (2014) ARA&A 52:415
70. Schiminovich D, Ilbert O, Arnouts S et al (2005) ApJ 619:L47
71. Le Floc'h E, Papovich C, Dole H et al (2005) ApJ 632:169
72. Reddy NA, Steidel CC (2009) ApJ 692:778
73. Smolčić V, Schinnerer E, Zamorani G et al (2009) ApJ 690:610

Chapter 8
Conclusion

In this chapter, we summarize our main results and discussions in the scales of the ISM, the CGM, and the cosmic structure.

Insterstellar Medium Scale

In Chap. 3, we analyse 1627 deep ALMA maps in Band 6 and 7 that became public by 2017 July and study the statistics of the rest-frame FIR effective radius ($R_{e(FIR)}$). The sample is composed of 1034 continuum sources detected at the submm and mm wavelengths, which is the largest sample ever made with the ALMA data. We identify 577 optical-NIR counterparts with the photometric redshifts at $z = 0 - 6$ and confirm from the redshift distribution that there exists no pre-selection bias in the initial ALMA observations. We measure the $R_{e(FIR)}$ values for the large sample based on the homogeneous uv-visibility size analyses with the exponential disk model ($n = 1$) Evaluating the We carefully evaluate the completeness in the source selection and the size measurement via Monte-Carlo simulations and address the $L_{FIR} - R_{e(FIR)}$ relation and the $R_{e(FIR)}$ evolution. We discuss the physical origins of the dusty starbursts by comparing our results with the rest-frame UV morphology and size. The main findings of this section are summarized below.

1. Our ALMA sources show the M_{star} values of $\sim 10^{10} - 10^{11.5}\,M_{\odot}$ and the SFR values of $\sim 100 - 1000\,M_{\odot}\mathrm{yr}^{-1}$. We find that the half are the high-mass end of the main sequence, while the rest of the half of our ALMA sources are the starbursts. The fraction of the starburst is decreased at high redshifts, which is consistent with the previous studies.
2. The redshift distribution is unchanged as a function of L_{FIR} nor between ALMA Band 6 and 7. We obtain the median redshift of $z_{med} = 2.36$, which is consistent with the previous studies.

S. Fujimoto, *Demographics of the Cold Universe with ALMA*, Springer Theses,
https://doi.org/10.1007/978-981-16-4979-0_8

3. Owing to the largest ALMA sample, we successfully derive the $R_{e(FIR)} - L_{FIR}$ relation in a wide redshift range of $z = 0 - 6$, and find a positive correlation at the >99% significance level.
4. We obtain the best-fit slope of $\alpha^{FIR} = 0.28 \pm 0.07$ via the power-law fitting of $R_{e(FIR)} \propto L_{FIR}^{\alpha^{IR}}$. We confirm that the best-fit slope is not affected by the contamination either of the lensed sources, the AGNs, or the potentially low-quality of early ALMA cycles. We also find the redshift evolution of $R_{e(FIR)}$: $R_{e(FIR)}$ decreases toward high redshifts at a fixed L_{FIR}. The best-fit α^{FIR} and the redshift evolution trend agree with the previous studies for the galaxy effective radius in the rest-frame UV ($R_{e(UV)}$) and optical wavelengths ($R_{e(Opt.)}$).
5. On the statistical basis, the average $R_{e(FIR)}$ values at $z = 1{-}6$ are smaller than those of $R_{e(Opt.)}$ and $R_{e(UV)}$ at a given M_{star} range. On the individual basis, the comparison between the $R_{e(FIR)}$ and $R_{e(Opt.)}$ ($R_{e(UV)}$) in individual galaxies also supports the trend of $R_{e(FIR)} \lesssim R_{e(Opt.)}$ ($R_{e(UV)}$). The dusty starbursts are suggested to be taking place in the compact regions by these statistical and individual results.
6. The median value of the spatial offset between the rest-frame UV-optical and FIR continuum is estimated to be 0.″24, which might be be explained by the astrometric uncertainty (∼0.″25) between ALMA and HST reported in the previous studies. There are some sources with the spatial offsets between the rest-frame UV-optical and FIR continuum larger than the astrometric uncertainty, indicative of the intrinsic spatial offsets between the rest-frame UV and FIR star-forming regions, instead of the misidentification for the optical-NIR counterparts due to the chance projection.
7. We test whether our ALMA sources are major mergers based on the rest-frame UV and optical morphology with the deep HST images. We identify 27% of our ALMA sources as the early/mid-stage major mergers, while the rest of the 73% as the isolated galaxies. The total fraction of the major mergers including the isolated galaxies with the merger-like morphology is estimated to be ∼30–40% in our ALMA sources. This indicates that the dusty starbursts are triggered not only by the major mergers but also the other mechanism(s).

In Chap. 4, we conduct the visibility-based stacking to the ALMA sources identified in the ASAGAO survey and further study the Sérsic index n and the effective radius R_e in the rest-frame FIR wavelength, n_{FIR} and $R_{e,FIR}$. The ALMA high-resolution (∼0.″19) 1.2-mm imaging of the ASAGAO survey covers the 26 arcmin^2 area in GOODS-S imaged with WFC3/IR on HST, which also enables us to obtain R_e and n in the rest-frame optical wavelength, $R_{e,opt}$ and n_{opt}. By using an additional sample of individual ALMA sources from the archive, we examine the morphology of the dusty star-forming galaxies in the rest-frame optical and FIR wavelengths as a function of L_{IR} over a wide range of $\sim 10^{11} - 10^{13}\, L_\odot$. We then discuss the evolutionary sequence from the high-z dusty star-forming galaxies to the compact quiescent galaxies. We summarize the major findings as follows.

7. The33 ASAGAO sources shows a 29σ level detection via the visibility-based stacking, which enables us to securely measure the average Sérsic profile. Carefully evaluating the smoothing effect due to the positional uncertainty via Monte-

Carlo simulations, we obtain the best-estimates of $R_{\rm e,FIR} = 1.0 \pm 0.2$ kpc and $n_{\rm FIR} = 1.2 \pm 0.2$.

8. By adding individual measurements for 12 bright ALMA sources in the archive, we evaluate $n_{\rm FIR}$ and $R_{\rm e,FIR}$ as a function of $L_{\rm IR}$ in the range of $\sim 10^{11} - 10^{13}\,L_{\odot}$. We find that the $n_{\rm FIR}$ measurements are unchanged from $n_{\rm FIR} = 1.2 \pm 0.2$ by $L_{\rm IR}$. On the other hand, the $R_{\rm e,FIR}$ measurements fall on the positive power-law correlation between $L_{\rm IR} - R_{\rm e,FIR}$ reported in the previous ALMA studies.
9. The distributions of $n_{\rm opt}$ and $R_{\rm e,opt}$ show no significant difference between the bright SMGs identified in the ALESS survey and the less bright ASAGAO sources. The statistical test results indicate that neither $n_{\rm opt}$ and $R_{\rm e,opt}$ depend strongly on $L_{\rm IR}$. We obtain the best-estimates of $R_{\rm e,opt} = 3.2 \pm 0.6$ kpc and $n_{\rm opt} = 0.9 \pm 0.3$ in the $L_{\rm IR}$ range of $\sim 10^{11} - 10^{13}\,L_{\odot}$.
10. Comparing the rest-frame optical and FIR results, we find that n always takes the common value of $\sim$1 regardless of the wavelengths. We also find $R_{\rm e,FIR} < R_{\rm e,opt}$, which agrees with the previous ALMA studies. These results suggest a general picture of the high-redshift star-forming galaxy: the dusty star formation takes place in a compact disk-like structure embedded in a larger stellar disk.
11. The best-fit Sérsic profiles in the rest-frame FIR+UV and optical provides us the radial surface density profile of $\rm SFR_{UV+FIR}$ and $M_{\rm star}$, respectively. Assuming the constant star-formation, our simple model reproduces the decreasing trend of $R_{\rm e,opt}$ from $z \sim 2$ dusty star-forming galaxies to $z \sim 1 - 2$ compact quiescent galaxies. However, the increasing trend of $n_{\rm opt}$ is not reproduced, where other mechanism(s) are required such as kinematic dissipation.
12. In the HR map of the stacked ASAGAO source, we find that there is a compact component at the center. The two-component fitting results show that the stacked profile is explained by a point source (PS)+disk profile rather than a bulge+disk profile. The PS component is likely to be originated by the central AGN, while one cannot rule out a possibility that there is a PS-like very compact dusty bulge at the center.

In Sect. 7.1, we discuss the physical origins of the ALMA submm/mm sources and the role in the galaxy evolution based on the above results.

13. We analyse the rest-frame UV and optical morphology with the deep HST images and test whether major mergers dominate in our ALMA sources. We find 73% of our ALMA sources classified as the isolated galaxies, while the rest of 27% as the early/mid-stage major mergers. When we regard the isolated galaxies with the merger-like morphology also as the major mergers, the fraction is increased to $\sim$30–40% in our ALMA sources. This indicates that the majority of the dusty starbursts are not the major merger system and that other mechanisms are required to trigger the dusty starbursts.
14. We study the detail morphology in the rest-frame optical wavelength for the $z \sim 1 - 3$ ALMA sources and find that on-going merger processes are likely important in high-z ULIRGs, but not in LIRGs. However, we caution the galaxies classified as the isolated disk are the most abundant even in the high-z ULIRGs,

which implies that the dusty star formation even in the ULIRGs at $z = 1 - 3$ is triggered not only by the on-going merger process.

15. On the size–stellar mass plane, $R_{e(opt.)}$ of the quiescent galaxies at $z = 1 - 3$ and $R_{e(FIR)}$ of our ALMA sources shows the similar distribution. This agrees with an evolutionary sequence from the high-z dust starburst to the local elliptical galaxies: the intense star-formation of dusty starbursts in compact regions forms the major part of the stellar distribution of the host that are evolved to the compact quiescent and the local elliptical galaxies.
16. From the best-fit Sérsic profiles in the rest-frame optical and UV+FIR, we derive the radial surface density profile of M_{star} and SFR_{UV+FIR}, respectively. The observed trend of $R_{e,opt}$ decreasing from dusty star-forming galaxies at $z \sim 2$ to compact quiescent galaxies at $z \sim 1 - 2$ is well reproduced from our simple model which assumes a constant star formation. However, the same model does not reproduce the increasing trend of n_{opt}. This indicates that the increasing trend of n_{opt} requires other mechanism(s) such as kinematic dissipation.
17. We find that there is a compact component at the center in the HR map of the stacked ASAGAO source. The two-component fitting results suggest that a point source (PS)+disk profile more likely explains the stacked profile rather than a bulge+disk profile. We interpret that the central PS component is attributed to the AGN, although we cannot rule out the possibility that a PS-like very compact dusty bulge is formed at the center.

Cicrumgalactic Medium Scale

In Chap. 5 and Sect. 7.2, we study the existence of the carbon-enriched gas halos around normal star-forming galaxies whose [CII] line have been individually detected at $z = 5.153 - 7.142$ via the visibility-based stacking with the deep ALMA data. Our stacking approach achieves the well-sampled deep visibility data in the uv-plane, which allows us to explore the diffuse emission spread over the circumgalactic scale. Also utilizing the deep HST/H-band data, we compare the radial profiles of the rest-frame UV and FIR continuum as well as the [CII] line emission. In Sect. 7.2, We discuss what physical origins form the extended [CII] line emission. We summarize the major findings as follows.

18. The visibility-based stacking of our and archival deep ALMA data produces 10σ and 21σ level detections at the peak for the dust continuum and the [CII] line emission, respectively, as average properties of 18 galaxies with SFR $\simeq$ $10–70 M_{\odot}$ at $z = 5.153 - 7.142$. The stacked [CII] line clearly shows its morphology extended more than the distribution of the dust continuum. In fact, the stacking results of the radial surface brightness profiles of the [CII] line are extended up to radius $\sim$10-kpc scale (9.2σ), which is consistent within the scatters of the individual results.

19. We also perform the rest-frame UV continuum stacking for 9 out of the 18 [CII] line sources whose deep HST/H-band data have been taken in the archive. In addition to the rest-frame FIR dust continuum, the radial surface brightness profiles of the [CII] line extends more than the rest-frame UV continuum emission. We examine the radial ratio of $L_{[\mathrm{CII}]}$/SFR$_{\mathrm{UV+FIR}}$ and find that the ratio becomes higher towards the halo areas, where the high ratios cannot be explained by the star-forming or dwarf galaxies. This indicates that the [C II] halo is not originated by the faint, low-mass satellite galaxies.
20. The two-component (Sérsic+exponential) profile fitting results indicate that the extended [CII] component has a scale length of 3.3 ± 0.1 kpc ($=5.6 \pm 0.1$ kpc in effective radius) that is comparable to the Lyα halo universally found around the high-z star-forming galaxies. Based on the comparison in the effective radius, the extended [CII] component has a larger radius by a factor of $\sim$5 than that of the central galactic component.
21. The state-of-art zoom-in simulation reproduces the radial surface brightness profile trends of the extended [CII] line emission and of the rest-frame UV comparable to the rest-frame FIR continuum emission. However, the zoom-in simulation does not reproduce a large part of the [CII] line luminosity in the extended component, where the zoom-in simulation may miss other mechanisms associated to the feedback process to form the extended [CII] halo component in early galaxies.
22. The existence of [C II] haloes around these early galaxies raises two questions: what powers the [C II] line emission in the CGM and how is the carbon abundance enriched at such an early cosmic epoch. Although there remain several possible scenarios that are attributed to the extended [C II] emission in the CGM scale, outflows are essentially required in all cases to enrich the carbon abundance of the primordial CGM gas around the early galaxies. Our results are thus the first evidence of outflow activities in these early galaxies.
23. The same stacking procedure is applied to another ALMA dataset for 34 QSOs at $z = 5.78$–7.09, which produces not only the [C II] halo but also the rest-frame FIR continuum halo extended over the radius of 10–15 kpc scale. The radial ratios of $L_{[\mathrm{CII}]}/\mathrm{SFR}_{\mathrm{IR}}$ cannot explain the extended [C II] and rest-frame FIR continuum, suggesting that the strong outflow (radiation) originated from the QSOs expels (heats) the dust into (in) the CGM and the thermal dust emission in the CGM is identified as the rest-frame FIR continuum halo.

Cosmic Structure Scale

In Chap. 6, we analyse one cluster, Abell 1689, and fifty blank-field maps taken by $\sim$120 deep ALMA Band 6 and 7 pointings and identify 133 faint ALMA continuum sources at 1.2 mm with a flux density of $0.02 - 1$ mJy. This sample explores the faintest regime in the 1.2 mm survey known to date reaching down to 0.02 mJy, which is drawn from the complete archive search for the deep ALMA data open for public by 2015 June and our deep ALMA projects. Using this faint 1.2-mm source sample, we derive the 1.2-mm number counts and estimate the contribution of the faint-source to the CIB. In Sect. 7.3, we also examine the physical origin of these faint 1.2-mm sources based on the statistical approach of clustering analysis as well as the individual approach of studying optical-NIR counterparts. We summarize the major findings as follows.

24. We derive the 1.2-mm number counts down to a flux density of 0.02 mJy by combining the faint and bright number counts from our and the previous studies, respectively. The number counts are consistent with the previous results using the single-dish lensing and ALMA data, and significantly push the flux density limit to the faintest regime. We find that the Schechter function well represents the number counts at $\gtrsim$0.02 mJy
25. Based on the derived number counts, we estimate the total integrated 1.2-mm flux density to be $22.9^{+6.7}_{-5.6}$ Jy deg^{-2} down to the flux limit of $S_{\rm limit}$ of 0.02 mJy. This corresponds to 104^{+31}_{-25}% of the CIB measured by *COBE* observations. Since the CIB is fully resolved down to ~0.02 mJy within the errors, the major contributors to the CIB at 1.2 mm are the individual sources with $\gtrsim$0.02 mJy. In other words, the contribution to the CIB of the 1.2-mm sources fainter than ~0.02 mJy should be negligibly small, suggesting a potential flattening and/or truncation in the number counts at the further faint flux density regime below ~0.02 mJy.
26. We analyse the clustering properties of our faint ALMA sources via the counts-in-cells method, assuming the sources redshift at $z \sim 1 - 4$. We constrain the galaxy bias with an upper limit of $b_g < 3.5$. This is smaller than those of massive galaxy populations of SMGs and DRGs, but comparable with those of UV-bright LBGs and sBzK populations.
27. We search for optical-NIR counterparts of our faint ALMA sources, focusing in two regions of Abell 1689 and SXDS that have rich multi-wavelength optical-NIR images. We identify 17 optical-NIR counterparts and find that 59% of our faint ALMA sources have the counterparts that are detectable in the optical-NIR continuum ($m \lesssim 28$ mag). The mid-IR diagnostics indicates that there are no AGN activities in these sources, which is confirmed from the absence of the X-ray or radio counterparts in any of our faint ALMA sources. We apply the color selection criteria of LBGs, BX/BM, pBzK, and sBzKs to our optical-NIR counterparts that have photometric data sets required for these color selections. We find that a majority of the optical-NIR counterparts meet color criteria of LBG, BX/BM, and sBzK, where the magnitudes are also comparable to these populations. Because we find that 59% of our faint ALMA sources are detectable in the optical-NIR band and classified as LBG, BX/BM, and sBZK populations, we conclude that about a half of faint ALMA sources are are high-z star-forming galaxies with the dust extinction and SFR much smaller than SMGs. Besides, the optical-NIR counterparts of the faint ALMA sources have the IRX-β relation clearly different from (U)LIRGs, indicating that the faint ALMA sources are not

explained by miniature (U)LIRGs that are simply scaled down with the infrared brightness.

28. We investigate our ALMA 1.2-mm number counts, comparing with the eleven 1.2-mm number counts estimated from the eleven LFs at $0 \le z \le 4$. We find that our ALMA and the total of the eleven 1.2-mm number counts are consistent within the uncertainties at $S_{1.2\,\mathrm{mm}} \ge 0.1$ mJy, and that the estimates for our ALMA 1.2-mm number counts are reliable. From the contribution fractions of these eleven to our ALMA 1.2-mm number counts, we confirm that the top hat assumption at $1 \le z \le 4$ agrees with our n(z) estimate for the faint ALMA sources. We also derive twelve 1.2-mm number counts from twelve SMFs at $0 \le z \le 8$. The total of the twelve 1.2-mm number counts, on the other hand, shows a deficiency against our ALMA 1.2-mm number counts at $S_{1.2\,\mathrm{mm}} \ge 0.4$ mJy beyond the uncertainties. Although the SMF analysis with another M_{star}–SFR relation partly fills the deficiency, the remaining deficiency implies the possibility that these are the dust obscured star-forming galaxies missing in the optical-NIR observations.

From Inter-stellar and Circum-Galactic to Cosmic Structure Scale

From the ISM and CGM to the cosmic structure scales, our results suggest that the dust obscured star-formation is taking place in the ISM scale, instead of widely distributed outside of the galaxies in the CGM and/or IGM scales, where we succeed in identifying the entire population of the dust obscured star-formation contributing to the CIB with ALMA.

29. We estimate the $\mathrm{SFRD_{total}}$ values at $1 \le z \le 4$, combining the results from the previous $\mathrm{SFRD_{UV}}$ studies and our 1.2-mm number counts, with the top hat assumption of n(z). We find that our $\mathrm{SFRD_{total}}$ estimate shows the excess to the average of the previous best-estimate $\mathrm{SFRD_{total}}$ value by 40^{+20}_{-40} %. Although this excess is only at 1σ significance level, this result supports the possibility that there exist the dust obscured star-forming galaxies missing in the optical-NIR observations.

Curriculum Vitae

Seiji Fujimoto
Cosmic DAWN Center, Niels Bohr Institute, University of Copenhagen
Jagtvej 128, DK-2200 Copenhagen N, Denmark
E-mail: fujimoto@nbi.ku.dk

Research Interests

- Evolution of massive galaxies
- Structure formation in the early universe
- Interplay among galaxy, circumgalactic and intergalactic media

Work Experience

- 2021–present EU co-fund INTERACTIONS Fellow, Cosmic DAWN Center
- 2019–present DAWN Fellow, Cosmic DAWN Center
- 2019–2019 ALMA Project Researcher, NAOJ/University of Waseda
- 2019–2019 ICRR Project Researcher, University of Tokyo
- 2016–2019 JSPS Research Fellow (DC1), University of Tokyo

S. Fujimoto, *Demographics of the Cold Universe with ALMA*, Springer Theses,
https://doi.org/10.1007/978-981-16-4979-0

Education

- 2016–2019 Ph.D Graduate school of Science, Department of Astronomy, University of Tokyo (Advisor: Prof. M. Ouchi)
- 2014–2016 M.Sc. Graduate school of Science, Department of Astronomy, University of Tokyo (Advisor Prof. M. Ouchi)
- 2010–2014 B.Sc. Department of Astronomy, University of Tokyo (Advisor: Prof. K. Kohno)

Awards/Prizes

- 2021 Inoue Research Award for Young Scientists
- 2019 University of Tokyo School of Science Research Award for Doctoral Students
- 2016 University of Tokyo School of Science Research Award for Master Students
- 2016 Institute for Cosmic Ray Research President's Award
- 2015 University of Tokyo President's Award for Students

Awarded Telescope Times (S. Fujimoto as a PI)

• JWST: Cycle 1 GO 1567 12h
• VLT: 108.22MK 30h
• Subaru: S16A0033N 15h, S20A0045N 10h S20B0150S 4h S21B0108N 18h, S22A0094N 27h
• ALMA: 2017.1.00531.S 18h, 2019. 1.00672.S 12h, 2019.1.00236.S 10h, 2019.2.00050.S 42h, 2021. 1.00055.S 16h, 2021.1.00247.S, 19h
• NOEMA: DDT E19AD 9h, E20EO 5 h, E20EN 2h, S21DM 34h, W21EH 26h, W21EF 2h
• JVLA: DDT 20A-520 13h, 21A-145 22h, 21A-162 23h
• JCMT: M17BP073 24h, M18AP001 30h
• SMA: 2020B-S051 24h

Outreach

- 2021 Press Release, "ALMA Discovers Rotating Infant Galaxy with Help of Natural Cosmic Telescope"
- 2019 Press Release, "Carbon Cocoon Surrounded Growing Galaxies—ALMA Spots Earlies Environment Pollution in the Universe—"

- 2016 Press Release, "ALMA Resolves the Cosmic Infrared Background Light"
- 2016–2018 Outreach public talk "Science Lab" in Hikawa High School, Japan
- 2017 Web Article "Beyond Connecting Dots", School of Science News in U.Tokyo

International Conferences in the Past 5 years

Summary: Oral talks in peer-reviewed and invited conference (**18**), other oral talks (**26**)

Highlights

- 11. 2022, Resolving Extragalactic Universe with ALMA, Tokyo, Japan (Invited)
- 03. 2022, The Growth of Galaxies in the Early Universe—VII, Sesto, Italy (Invited)
- 03. 2022, Linking Galaxy Physics from ISM to IGM Scales, Sesto, Italy (Invited)
- 06. 2021, Summer All Zoom Epoch of Reionization Astronomy Conference 2, online
- 02. 2021, CIDER—The Cold ISM During the Epoch of Reionization, online
- 10. 2020, The rise of metals and dust in the Universe over a wide range of cosmic times, online
- 07. 2020, Summer All Zoom Epoch of Reionization Astronomy Conference, online
- 11. 2019, Ringberg Workshop, Ringberg, Germany (Invited)
- 11. 2019, Revolutionary Spectroscopy of Today as Springboard to Webb, Leiden, Netherlands (Invited)
- 10. 2019, ALMA 2019: Science Results and Cross-Facility Synergies, Cagliari, Italy
- 09. 2019, Views on the ISM in galaxies in the ALMA era, Bologna, Italy -
- 09. 2019, Extremely Big Eyes on the Early Universe, Roma, Italy -
- 09. 2018, Chili-Japan Academic Forum, Nikko, Japan (Invited) -
- 08. 2017, Twenty years of Submillimeter Galaxies, Durham, England